W0042280

METHODS IN MEMBRANE BIOLOGY

VOLUME 5
Transport

Contributors to This Volume

Diana Dowd, *Worcester Foundation for Experimental Biology, Shrewsbury, Massachusetts*

Joy Hochstadt, *Worcester Foundation for Experimental Biology, Shrewsbury, Massachusetts*

E. R. Kashket, *Department of Microbiology, Boston University School of Medicine, Boston, Massachusetts*

George A. Kimmich, *Department of Radiation Biology and Biophysics, School of Medicine and Dentistry, University of Rochester, Rochester, New York*

Chien-Chung Li, *Worcester Foundation for Experimental Biology, Shrewsbury, Massachusetts*

Peter C. Maloney, *Department of Physiology, Harvard Medical School, Boston, Massachusetts*

Dennis C. Quinlan, *Worcester Foundation for Experimental Biology, Shrewsbury, Massachusetts*

Richard L. Rader, *Worcester Foundation for Experimental Biology, Shrewsbury, Massachusetts*

Kenneth Sisco, *Laboratory of Neurobiology, National Institute of Mental Health, Bethesda, Maryland*

Ichiji Tasaki, *Laboratory of Neurobiology, National Institute of Mental Health, Bethesda, Maryland*

T. H. Wilson, *Department of Physiology, Harvard Medical School, Boston, Massachusetts*

A Continuation Order Plan is available for this series. A continuation order will bring delivery of each new volume immediately upon publication. Volumes are billed only upon actual shipment. For further information please contact the publisher.

METHODS IN MEMBRANE BIOLOGY

VOLUME 5
Transport

Edited by EDWARD D. KORN

Laboratory of Cell Biology
National Heart and Lung Institute
Bethesda, Maryland

PLENUM PRESS • NEW YORK AND LONDON

Library of Congress Cataloging in Publication Data

Korn, Edward D   1928-
    Methods in membrane biology.

    Vol. 4 has title: Biophysical approaches; v. 5, Transport.
    Includes bibliographies.
    1. Membranes (Biology) I. Title. [DNLM: 1. Membranes—Periodicals. W1
ME9616C]
QH601.K67   574.8'75   73-81094

ISBN 978-1-4684-6978-3   ISBN 978-1-4684-6976-9 (eBook)
DOI 10.1007/978-1-4684-6976-9

© 1975 Plenum Press, New York
Softcover reprint of the hardcover 1st edition 1975
A Division of Plenum Publishing Corporation
227 West 17th Street, New York, N.Y. 10011

United Kingdom edition published by Plenum Press, London
A Division of Plenum Publishing Company, Ltd.
Davis House (4th Floor), 8 Scrubs Lane, Harlesden, London, NW10 6SE, England

All rights reserved

No part of this book may be reproduced, stored in a retrieval system, or transmitted,
in any form or by any means, electronic, mechanical, photocopying, microfilming,
recording, or otherwise, without written permission from the Publisher

Articles Planned for Future Volumes

Cell Fractionation Techniques
   H. Beaufay and A. Amer-Costesec (Université Catholique de Louvain)

Methods of Isolation and Characterization of Bacterial Membranes
   M. R. J. Salton (New York University Medical Center)

Techniques of Immunoelectron Microscopy and Immunofluorescence in the Study of Membrane Components
   E. de Petris (Basel Institute for Immunology)

Synthesis of Stereospecific Phospholipids for Use in Membrane Studies
   M. Kates (University of Ottawa)

Affinity Chromatography in Membrane Research
   P. Cuatrecasas (Johns Hopkins University)

Electron Microscopy of Membranes
   H. P. Zingsheim and H. Plattner (Max-Planck Institut für Biophysikalische Chemie and University of Munich)

Determination of Asymmetric Phospholipid Distribution in Membranes
   R. F. A. Zwaal and B. Roelofsen (University of Utrecht)

Isolation and Characterization of Membrane Binding Proteins
   D. L. Oxender and S. C. Quay (University of Michigan)

Selection and Study of Bacterial Mutants Defective in Membrane Lipid Biosynthesis
   D. F. Silbert (Washington University)

Scanning Calorimetry of Membranes and Model Membranes: Theory and Data Interpretation
   J. M. Sturtevant (Yale University)

Electron Spin Resonance Studies of Membranes
   B. Gaffney (Johns Hopkins University)

Methods of Reconstruction of Transport
   P. C. Hinkle (Cornell University)

X-Ray and Neutron Diffraction Studies of Membranes
   D. A. Kirschner, D. L. D. Caspar and L. Makowski (Brandeis University)

Recent Methods for the Structural Identification of Lipids
   R. Klein and P. Kemp (Cambridge University and ARC Institute of Animal Physiology)

Isolation and Characterization of Acetylcholine Receptors
   M. A. Raftery (California Institute of Technology)

Chemical Relaxation Spectrometry for the Investigation of Mechanisms
Involved in Membrane Processes
E. Grell (Max-Planck Institut fur Biophysikalische Chemie)

The Use of Organic Solvents in Membrane Research
P. Zahler (Universität Bern)

Lipid Exchange Between Membranes
D. B. Zilversmit (Cornell University)

Procedures for Labeling Surface Carbohydrates
S. Hakomori (University of Washington)

Mammalian Cell Membrane Mutants
R. M. Baker and V. Ling (Massachusetts Institute of Technology and
Ontario Cancer Institute)

Methods for Determining the Topographical Distribution of Proteins in Membranes
M. Morrison (St. Jude Children's Research Hospital)

Contents of Earlier Volumes

Preface

One property common to all cells is transport. Molecules and ions must enter and leave cells by crossing membranes in a controlled manner. The process may take any of several forms: simple diffusion, carrier-mediated diffusion, active transport, or group translocation. There is more than one way to measure each. Transport kinetics, with particular reference to the red blood cell, were discussed in a previous volume. Three chapters deal with the general subject of transport in this volume. Maloney, Kashket, and Wilson summarize the appropriate methodology for studying metabolite and ion transport in bacteria, and Kimmich describes the relevant methodology for the isolated intestinal epithelial cell. The methods described in these two chapters have general application to transport studies in single cells from any source.

The approach described in these two complementary articles is extended in the chapter by Hochstadt and her collaborators on the use of isolated membranes from bacterial and mammalian cells for the study of transport phenomena. If one can prepare a suitable plasma membrane fraction (sealed, impermeable vesicles with the necessary transport components intact), it becomes possible to separate the events of transport from any subsequent metabolism that may occur in the cell. Isolated membrane vesicles are relatively easy to obtain from bacteria, and they are comparatively well studied. Work with similar preparations from cultured mammalian cells is just beginning but has much promise.

The advantages of these simplified systems are obvious, but it is always necessary to interpret the data in accordance with the properties of the intact cell. One component that may frequently be missing from the isolated membrane is the binding protein specific for the metabolite under study. In a future volume the isolation and characterization of transport binding proteins and their recombination into vesicles will be discussed.

There are also other types of transport that will be treated in future volumes: transport across intracellular membranes, for example, and transport of membrane constituents themselves.

A very specific form of transport, electrical conductance in excitable tissue, is the subject of the final chapter in this volume. Electrophysiology has contributed to, and profited from, recent developments in membrane biology and Tasaki has contributed vital theoretical and methodological insights in this field. In this volume, Tasaki and Sisco describe in detail the procedures for making electrical measurements in *Nitella*, which, as they point out, is an underused experimental model with electrical properties essentially identical to those of the nerve axon but without many of the experimental difficulties found in nerve research. These authors also describe their procedures for isolating single nerve fibers from frogs, crabs, and lobsters and for isolating and perfusing the giant squid axon. For about 5 years nerve physiologists have been developing optical methods (fluorescence and birefringence) for measuring phenomena associated with conductance. Special problems are created by the very low signal-to-noise ratio, and these problems require very special solutions that are detailed in this chapter.

Bethesda   Edward D. Korn

Contents

Chapter 2

Preparation and Characterization of Isolated Intestinal Epithelial Cells and Their Use in Studying Intestinal Transport

GEORGE A. KIMMICH

Chapter 3

Use of Isolated Membrane Vesicles in Transport Studies

JOY HOCHSTADT, DENNIS C. QUINLAN, RICHARD L. RADER, CHIEN-CHUNG LI, and DIANA DOWD

Chapter 4

Electrophysiological and Optical Methods for Studying the Excitability of the Nerve Membrane

ICHIJI TASAKI and KENNETH SISCO

Chapter 1

Methods for Studying Transport in Bacteria

PETER C. MALONEY, E. R. KASHKET,*
and T. H. WILSON

Department of Physiology
Harvard Medical School
Boston, Massachusetts

1. INTRODUCTION

The major objective of this chapter is to outline some of the important techniques which have been developed for the study of transport systems in bacteria. The techniques described are discussed in detail with respect to the study of the lactose transport system of *Escherichia coli*. It now appears that this transport system can serve as a useful model for a number of active transport systems in both bacterial and animal cells. In such transport systems, a protein (the "carrier") embedded within the membrane mediates the translocation and accumulation of substrate; substrate appears within the cell without chemical modifications. The driving force for the accumulation of substrate is represented by the electrical and chemical forces acting on certain specific cations. In bacterial cells, accumulation by such transport systems is coupled to proton movements (see discussions by Mitchell, 1963, and Harold, 1972), whereas in animal cells such active transport is associated with the movement of sodium ions (for review, see Schultz and Curran, 1970). In the absence of energy coupling, however,

* Present address: Department of Microbiology, Boston University School of Medicine, Boston, Massachusetts.

these systems catalyze the facilitated diffusion of substrate across the cell membrane.

This chapter focuses on several of the experimental approaches employed to characterize such transport systems in bacteria. Section 2 begins with an outline of some general methodology and then continues with a more extensive discussion of the techniques used to describe the kinetic parameters of the lactose transport system under conditions where substrate is accumulated within the cell. Section 3 maintains an emphasis on the lactose transport system but considers the methods employed to study transport in the absence of substrate accumulation. There is now strong evidence to suggest that certain bacterial transport systems, such as the lactose transport system of *E. coli* and the galactose transport system of *Streptococcus lactis*, couple the movement of substrate to that of protons. Thus accumulation of substrate occurs in the presence of chemical and electrical forces tending to move protons into the cell. Section 4 presents several techniques which may be used to examine each of these components of the "protonmotive force" (Mitchell, 1963). Methods are now available for the measurement of both the transmembrane pH gradient and the electrical potential across the membrane. Finally, Section 5 outlines less direct methods for studying membrane transport. These are methods which monitor the volume changes (swelling) that accompany the movement of substrate into the bacterial cell. These techniques have not been used extensively in the field of bacterial transport but can give valuable information when alternative methods are not available.

While specific examples given in this chapter are taken largely from work done in this laboratory, the methods they illustrate are generally applicable to other types of transport systems in microorganisms.

2. TECHNIQUES FOR MEASURING ACCUMULATION OF SUBSTRATE

2.1. Measurement of Intracellular Water

It is common practice in the field of bacterial transport to express the accumulation of substrate as in a biochemical reaction, e.g., moles of substrate taken up per gram of cellular protein. For transport processes, however, it seems more appropriate to express data in a way which focuses on the relationship between the intracellular and extracellular concentrations of substrate. It is this parameter, for example, which distinguishes a facilitated diffusion system (no intracellular accumulation of substrate) from an active transport system (accumulation of substrate). Mitchell and Moyle (1956) have noted that there is a high osmotic pressure within bacterial cells (about 20–25 atm). This implies that most low molecular weight compounds within the cell exist in free solution. The work of Sistrom (1958) clearly showed that the substrates accumulated by the lactose transport system in *E. coli* are in free solution within the cell and not bound to internal components. Thus the appearance of transport substrates within the cell is properly expressed in terms of their actual concentration. By using this convention one more readily appreciates that there need not be a fundamental difference in the interaction of substrate with the membrane carrier on either side of the membrane. Both processes will display concentration dependence, specificity, competition, etc.

To estimate the actual internal concentration of substrate, one must first measure the intracellular water volume. This is most conveniently determined from the difference between the wet weight and dry weight of cells, as described by Winkler and Wilson (1966). A concentrated cell suspension, in medium containing [³H]inulin (a marker for extracellular water), is placed into a tared centrifuge tube. After centrifugation and aspiration of the supernatant, the wet weight of the cell pellet (including trapped extracellular fluid) is determined. To minimize carryover of extracellular fluid, the sides of the centrifuge tube should be wiped free of adhering drops. Water is then removed by evaporation, either in a drying oven or in a desiccator over phosphorous pentoxide, and the dry weight is obtained. The difference between the dry and wet weights represents the weight of intracellular water plus the weight of trapped, extracellular water. The weight of extracellular water is accurately measured by resuspending the dry pellet and determining the amount of radioactive inulin present. It should be noted that different values for intracellular water volumes are obtained when different bacteria are examined. *E. coli* contains 2.7 µl of cell water per milligram dry weight (Winkler and Wilson, 1966; West, 1970), whereas *S. lactis* has 1.5 µl cell water per milligram dry weight (Kashket and Wilson, 1972a). Thus it is important to determine intracellular water volumes directly for each organism under study.

This procedure employs inulin as a marker of the extracellular water trapped in the cell pellet. Since inulin is too large to penetrate the cell wall of bacteria, cell water measured by this method represents both the water present in the cell and any water present in the periplasmic space between the cell wall and the plasma membrane. Under most experimental conditions, the volume of the periplasmic space is probably small compared to the true intracellular water volume. The osmotic pressure within bacterial cells is high and one expects that turgor pressure would keep the plasma membrane closely packed against the cell wall, limiting the size of the periplasmic space. An alternative procedure which avoids these considerations would be one in which the extracellular space is monitored by the distribution of a compound small enough to penetrate the cell wall but excluded by the cell membrane. D-Sorbitol does not penetrate the cell membrane in streptococci (although it is rapidly metabolized by *E. coli*) but is small enough to penetrate the cell wall. For *S. lactis*, only small differences were observed when cell water determinations were made with inulin or D-sorbitol as a marker for extracellular water (Kashket and Wilson, unpublished observations).

2.2. Assay Conditions and Separation of Cells from the Medium

Optimal conditions for active transport in bacteria are usually obtained using cells harvested from the exponential phase of growth. The preparation of stock cell suspensions will vary according to the experimental plan. In this laboratory, cells are usually washed, at 4°C, using the medium 63 (Cohen and Rickenberg, 1956) in which they were grown but with chloramphenicol (50 μg/ml) present and without added carbon source. After resuspension in a small volume of this salts medium, stock cells are maintained at 4°C until use. It is probably wisest to resuspend cells in the salts medium which supports growth, at least until the ion dependence of the system has been described. The pH of the resuspension and assay fluids is usually near neutrality.

For assay, cells from the stock suspension are diluted to the appropriate cell density in medium 63. For convenience, in this laboratory most studies are carried out at room temperature. However, it should be appreciated that the characteristics of a transport system show significant variation with temperature. Efflux experiments are usually done at low temperature (around 10°C) to take advantage of the reduced rates of exit at low temperatures. When conclusions depend primarily on a comparison of results obtained at different temperatures, the properties of the transport system should be verified at each temperature. For example, it is now known that at low temperatures there is an increase in the affinity of the lactose membrane carrier for its substrates (Sullivan *et al.*, 1974; Wilson, unpublished results).

Many transport systems in *E. coli*, including the lactose transport system, are sensitive to sulfhydryl reagents and for this reason β-mercaptoethanol or a similar reducing compound is often included in the assay mixture. Carter *et al.* (1968) have found that this prevents autooxidation of a sensitive sulfhydryl group on the lactose membrane carrier. In addition, a carbon source is sometimes included in the assay mixture (Kepes, 1960, 1971; Koch, 1963; Ganesan and Rotman, 1965). However, washed cells of *E. coli* contain sufficient endogenous reserves of "energy"-yielding materials to support active transport and special procedures must be devised to reduce these reserves and abolish accumulation (Koch, 1971; Berger, 1973). This is in contrast to the properties of organisms such as *S. lactis*, in which active transport is strictly dependent on the addition of fermentable compounds (Kashket and Wilson, 1972*a*).

After temperature equilibration of cells in the assay mixture, radioactively labeled substrate is added to initiate the reaction. So as not to grossly alter the osmotic composition of the reaction mixture, the substrate (and other additions) should be prepared in the appropriate salts medium or added as a small volume of an aqueous solution.

Separation of the cells from the medium is almost always done by filtration of cell suspensions through filters of small pore size (≤0.6 μm). This technique, first introduced by Britten *et al.* (1955) and Kepes (1960), offers considerable simplification over earlier methods which depended on a centrifugation step to separate cells from the medium (Rickenberg *et al.*, 1956). The filtration technique permits rapid sampling times, allowing accurate estimates of initial rates of both influx and efflux of substrate. Most investigators have adopted their own standard procedure for routine use. Since washing inevitably leads to some loss of intracellular substrate, some laboratories prefer not to wash and to use cells in which transport function has been blocked as a baseline (Rotman and Guzman, 1961). However, this is not sufficient for detailed kinetic studies, since the contaminating extracellular fluid on the filter may not be constant from sample to sample. Although one may monitor extracellular fluid on unwashed filters by including radioactive inulin in the assay mixture (Kashket and Wilson, 1973), some sort of washing procedure is usually employed. If the sample (160–220 μg dry weight of cells, equivalent to 0.4–0.6 μl of cell water for *E. coli*) is placed in the center of a presoaked filter, taking care not to allow fluid to be trapped underneath the chimney, one wash with 5 ml of medium 63 (room temperature) removes essentially all of the extracellular substrate while limiting the loss of internal substrate to less than 5%. With practice, the time between the beginning of sample filtration and the completion of the wash can be as little as 6 s.

The composition and temperature of the wash fluid must be chosen with care. Often, the wash is done with iced medium to restrict efflux of substrate after sampling (Kepes, 1960, 1971; Koch, 1963). However, in his careful study of the filtration technique, Leder (1972) observed that if the temperature of the wash fluid was less than 10°C there was a rapid loss of internal substrate unless the wash fluid was hyperosmotic with respect to the assay mixture. This was found for both amino acid and sugar transport in *E. coli*. Since Leder found that different strains (or the same strain grown under different conditions) exhibited varying susceptibility to this cold shock, it is recommended that a temperature transition during the wash be avoided. Kepes (1971) noted that the cold shock during washing accounted for the dramatic inhibition of thiomethylgalactoside accumulation by *E. coli* when cells were incubated with 0.2 M NaCl during the assay. Others have also observed that internal substrates may be lost when cells are subjected to a cold shock (Britten and McClure, 1962; Ring, 1965; Novotny and Englesberg, 1966).

2.3. Determination of Kinetic Parameters

Of the transport systems in bacteria, the lactose transport system of *E. coli* has been the most thoroughly studied with respect to the kinetic aspects of carrier-mediated substrate movements. Thus this section deals primarily with methods which have been used to characterize active transport by this system.

At least three components must be considered in a quantitative treatment of the lactose transport system: (1) carrier-mediated influx, (2) carrier-mediated efflux, and (3) diffusion pathways. Each of these may be evaluated by the appropriate techniques.

2.3.1. Carrier-Mediated Influx

In most instances, measurement of the initial rate of entry of substrate does not present serious technical difficulties. Rapid sampling using the filtration technique allows 4 or 5 points to be taken during the first 60 s, and this is usually sufficient to determine the interval during which the rate of entry is linear with time. Unless it has been established that the rate of entry has a 0 time intercept, one should not rely on single time points as a measure of initial rates.

Even when samples are taken as early as possible, the internal concentration of substrate is often significantly higher than that of the external medium at early times. Such measurements may underestimate the true initial rate if significant efflux of substrate has occurred. In an attempt to control for this possibility, one may examine the entry of labeled substrate into cells preloaded with unlabeled substrate. Unlabeled molecules will accumulate within the cell during the preincubation period and, because of their high internal concentration, will block efflux of newly entering radioactive molecules. Using this procedure, Kepes (1960) and Winkler and Wilson (1966) showed that for the lactose transport system the initial rate of entry of substrate was not significantly affected by preloading. (Under different experimental conditions, preloading can alter the observed rates of entry. This phenomenon, termed "exchange diffusion," will be discussed in a later section.)

Preloading techniques have been especially useful for the study of the lactose transport system in energy-poisoned cells. In such cells the affinity of the carrier for the efflux step is markedly increased (Winkler and Wilson, 1966). Consequently, rates of *entry* can be seriously underestimated because efflux becomes appreciable even at low concentrations of substrate. In this case, preloaded cells must be used to obtain accurate estimates of initial rates of influx. This is done by incubating cells at room temperature in the presence of a high concentration of substrate (about 20 times the estimated K_m for efflux), removing external substrate by centrifugation and washing at 4°C, and finally resuspending cells in a small volume of iced medium. Aliquots of this concentrated shock suspension are then diluted in assay medium containing radioactive substrate at the desired concentration and samples are rapidly taken for initial rate measurements. This technique should be applicable to any transport system in which the K_m for efflux is close to the K_m for entry. The slow chilling of cells during centrifugation does not result in efflux of substrate, as does the rapid cold shock which occurs during filtration and washing with iced medium (Leder, 1972). One might imagine that the "warm" shock involved in this procedure might present problems. However, Winkler and Wilson (1966) showed that both the K_m and the V_{max} for entry in preloaded, energy-poisoned cells were the same as those measured for entry into untreated cells (see Section 3.1).

2.3.2. Carrier-Mediated Efflux

Accurate measurements of initial rates of efflux are difficult to obtain, since one attempts to estimate changes in internal concentration which are small compared to the initial level. The usual procedure is to incubate a concentrated suspension of cells with labeled substrate under accumulating conditions; the final internal level of substrate is varied by using different external concentrations. After accumulation of substrate, the cells are centrifuged in the cold (to restrict efflux) into a compact pellet, the supernatant is decanted, and the sides of the centrifuged tube are wiped free of adhering fluid. It is usually not necessary to wash cells after centrifugation. The cells are then diluted (at least 20-fold) by adding warm medium without substrate. If care is taken to wipe the tube free of fluid, samples need not be washed after filtration since a large part of the radioactivity is contained within cells and counts in the medium are effectively removed by filtration alone. For rapid filtration, sample volume should be equivalent to about 250 μg dry weight of cells or less. As in the case of influx measurements, it is important to use multiple points to establish the rate of efflux.

In efflux studies it is particularly important to evaluate the nonspecific, "diffusion" component of exit by examining cells in which carrier activity is blocked. In this regard, it should be pointed out that conditions which block accumulation do *not* necessarily block carrier activity. Under most conditions, metabolic inhibitors (azide, dinitrophenol, etc.) which block accumulation of substrates by the lactose transport system do not significantly reduce the capacity of the carrier to mediate either the entrance or the exit of substrate (Winkler and Wilson, 1966). One must use cells in which the carrier is not present (mutants, or uninduced cells if their basal level of carriers is low) or in which the carrier has been truly inhibited (e.g. by sulfhydryl reagents in sensitive systems).

Efflux studies carried out as described above may be open to one serious criticism. This is illustrated by a simple example taken from the lactose transport system (Winkler and Wilson, 1966; Robbie and Wilson, 1969). At the steady state of accumulation, the rate of entry of substrate must be equal to its rate of exit. The rate of entry is readily measured from the initial kinetics of accumulation. However, if one takes the same cells, washes them free of medium, and resuspends them in identical medium without substrate, the rate of efflux is only about 10–15% of that determined for entry. The most probable explanation for this is a recapture phenomenon. That is, within the periplasmic space substrate molecules which have left the cell do not readily mix with the bulk of the extracellular fluid. Thus, after efflux, substrate molecules in this space have a high probability of being recaptured by the carrier and net efflux from the cell is significantly reduced. Experimentally, the solution to the recapture problem is analogous to the preloading experiments discussed above. One performs efflux measurements in medium containing sufficient unlabeled substrate (about 20 times the effective K_m for entrance) to block recapture of newly lost radioactive molecules (Fig. 1). Chilled cells containing labeled substrate are diluted

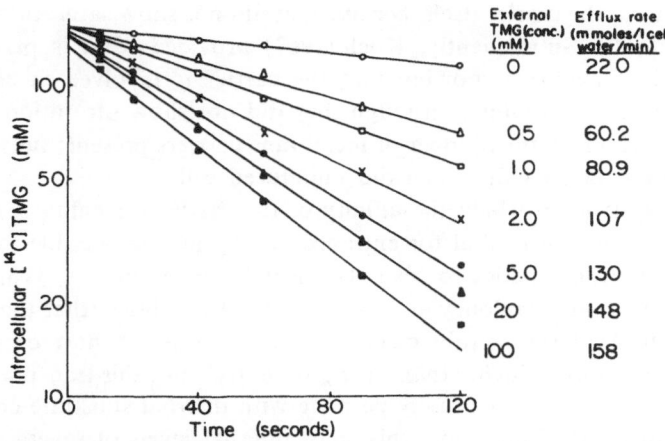

Fig. 1. Effect of external substrate on galactoside efflux. Cells of *E. coli* ML 308-831 (constitutive for the lactose transport system but lacking thiogalactoside transacetylase) were grown to the exponential phase in medium 63 containing 1% of a casein tryptone digest. After harvesting, cells were washed and resuspended in medium 63 containing 0.1 mg/ml chloramphenicol and 1 mg/ml glucose as an energy source. Cells (about 1.3 mg dry weight/ml) were equilibrated at 23°C with 20 mM [14C]thiomethylgalactoside (TMG) and then diluted 400-fold into media containing glucose plus various concentrations of nonradioactive substrate. Samples of 4 ml were filtered and washed at the indicated times. The concentration given to the right of each curve is that of the nonradioactive thiomethylgalactoside present in the efflux medium; the rate given is the corresponding initial rate of efflux, calculated from the kinetics of loss of radioactive substrate from the cells. Note that the ordinate is a logarithmic scale. From Robbie and Wilson (1969).

into assay medium containing unlabeled substrate rather than into substrate-free medium. When efflux experiments are performed in this way, exit rates can be shown to be approximately equal to entry rates measured under comparable conditions (Robbie and Wilson, 1969).

The general problem which plagues efflux studies is that exit from the cell is rapid, especially when recapture is blocked. Thus the internal substrate concentration may change markedly during the first 60 s, making it difficult to determine initial rates of exit as a function of substrate concentration. For these kinds of experiments, one generally uses low temperatures (around 10°C) to reduce the rate of exit. Even when all these precautions have been taken, however, efflux measurements may prove difficult to interpret, and in evaluating the role of the membrane carrier in exit one must not argue from kinetic studies alone. For example, early work on the lactose transport system suggested that efflux of substrate did not occur via the carrier itself, because exit did not show saturation kinetics (Kepes, 1960). Subsequently, Koch (1963) provided what is perhaps the best kind of evidence showing that the carrier is involved in efflux. He found that rates of efflux, although they did not show saturation kinetics, were dependent on the number of membrane carriers present; fully induced cells showed faster efflux than did uninduced cells.

For systems in which the affinity of the carrier for efflux is normally very low compared to that for entrance, it may not be possible to demonstrate saturation kinetics for exit. Using indirect arguments, Winkler and Wilson (1966) and Maloney and Wilson (1973) concluded that the effective K_m for efflux of the lactose membrane carrier was 100 mM or above (2 orders of magnitude higher than the K_m for entry). For this transport system, efflux experiments must usually be done with internal substrate concentrations significantly lower than this since internal levels of several hundred millimolar are difficult to obtain. Consequently, rates of exit do not show a clear hyperbolic (saturation) dependence on substrate concentration. In addition, the rate of efflux from the cell is usually found to be exponential with time (as in Fig. 1). Both of these observations are to be expected if the true K_m for efflux is greater than the substrate levels actually used, and such findings do not prove that exit is via "diffusion" or other nonspecific pathways. It is because of these kinds of problems that there has as yet been no satisfactory measurement of the affinity of the lactose membrane carrier for efflux in normal cells. However, in energy-poisoned cells, in which the carrier shows a marked increase in its affinity for the efflux step, the methods described above have proved sufficient to determine the K_m for exit (Winkler and Wilson, 1966).

2.3.3. Diffusion Pathways

As used here, the term "diffusion" applies both to free diffusion of substrate through the cell membrane and to the movement of substrate through unidentified carrier pathways which show diffusion-like kinetics under the conditions employed. The "diffusion" pathway must be evaluated from measurements of influx or efflux of substrate in cells which have no membrane carrier activity. For example, one might examine transport-negative mutants. Maloney and Wilson (1973) estimated the "diffusion" pathway for thiomethylgalactoside from the kinetics of entry and equilibration into cells which had not been induced for the lactose transport system and which, in addition, had been exposed to the competitive inhibitor thiodigalactoside to block residual transport by the basal level of carrier activity. Entry through the "diffusion" pathway was sufficiently slow so that standard methods for determining influx rates could be used. For smaller and less hydrophilic molecules, this may not be possible. Winkler and Wilson (1966) were able to evaluate the "diffusion" component for lactose flux by studying the efflux of lactose from cells whose transport function was blocked by prior exposure to p-chloromercuribenzoate. For both thiomethylgalactoside and lactose "diffusion" it could be shown that the initial rates of entry or efflux were directly proportional to substrate concentration over the range tested. For a 1 mM gradient of thiomethylgalactoside, the rate of entry via the "diffusion" route was equivalent to 0.14 μmol TMG/μl cell water per minute; for lactose efflux, the comparable figure was at least 10-fold lower. This difference is to be expected. Lactose is a disaccharide and by virtue of its larger size and greater number of hydrophilic groups would be expected to traverse the cell membrane less readily than thiomethylgalactoside. As will be discussed in the next section, the presence of a "diffusion" pathway for substrate movement plays an important role in determining the capacity of an active transport system to accumulate substrate.

2.4. The Steady State

In the steady state of accumulation, the internal concentration of substrate is maintained at a level higher than that of the external medium. Several factors contribute to the size of the gradient generated by this active transport. Among the most important of these are (1) the kinetic parameters (substrate affinity and velocity terms) describing the behavior of the carrier with respect to the entry and exit steps, (2) the number of membrane carriers, and (3) alternate, "diffusion" routes available for the movement of substrate. To illustrate how these various components influence the level of accumulation, it is useful to consider the steady state as in the following equation:

$$V_{max}^{influx} \frac{S_o}{K_m^{influx} + S_o} + DS_o = V_{max}^{efflux} \frac{S_i}{K_m^{efflux} + S_i} + DS_i$$

This equation simply states that, at the steady state, the rate of entry of substrate equals the rate of its exit.* Terms describing entry are given on

---

* Exchange diffusion occurs in bacterial transport systems (see Section 2.5). Including terms to accommodate this phenomenon does not alter the final form of the steady-state equation.

the left; terms describing exit are given on the right. The properties of the membrane carrier are given by Michaelis–Menten terms: the maximal velocities for entry (V_{max}^{influx}) and for exit (V_{max}^{efflux}) and the affinities of the carrier for the entry (K_m^{influx}) and exit (K_m^{efflux}) steps. S_o and S_i refer to the external and internal substrate concentrations, respectively, and D is a coefficient describing movement of substrate through a "diffusion" pathway.

According to this view, there are three simple mechanisms which would account for the accumulation of substrate: (1) the affinity of the carrier for efflux being less than that for entrance, (2) the maximal velocity for efflux being less than that for entry, or (3) some combination of these two. For the lactose transport system, the available information suggests that the first mechanism is the correct one. Under accumulating conditions, there is no difference between the maximal velocities for entrance and exit (Robbie and Wilson, 1969) whereas there is a large difference with respect to the affinity of the carrier for entry and exit (Winkler and Wilson, 1966).

Simplification of the steady-state equation shows that, in the absence of "diffusion" components, the gradient S_i/S_o would be determined solely by the ratio $K_m^{efflux}/K_m^{influx}$. This implies that the same steady-state gradient would be attained in cells containing different numbers of membrane carriers (different values for the V_{max} terms). However, for the lactose transport system it is known that experimentally the ratio S_i/S_o shows a striking dependence on the number of membrane carriers that are present (Koch, 1963; Maloney and Wilson, 1973). For example, in cells with 3% of the normal number of membrane carriers the steady-state gradient attained with thiomethylgalactoside is equivalent to only 15% of that found in normal cells. This is because of the presence of a "diffusion" component for transport. The contribution of "diffusion" to entry is usually minimal since low external concentrations of substrate may be used. However, in an actively accumulating system the *internal* concentration of substrate is often high and "diffusion" may account for a substantial fraction of efflux, especially when the number of membrane carriers is low. Using the steady-state equation given above, one may calculate the fraction of thiomethylgalactoside efflux proceeding via the "diffusion" pathway as a function of the number of membrane carriers present (varying values of V_{max} for the system). This relationship is shown in Fig. 2 and serves to illustrate some of the problems which may be encountered in interpreting steady-state measurements. When only a few carriers are present, most of the internal substrate leaves the cell via the "diffusion" pathway. In this region, the steady-state level of thiomethylgalactoside is nearly linearly proportional

Fig. 2. Calculated efflux by "diffusion" at the steady state. For thiomethylgalac- toside as a substrate, the steady-state equation (see text) was used to calculate the fraction of total efflux proceeding via the "diffusion" pathway in cells with different numbers of membrane carriers. Values for D (0.14 mmol/liter cell water/ min for a 1 mM gradient of substrate), K_m^{influx} (0.8 mM), and $V_{\text{max}}^{\text{influx}}$ (about 100 mmol/liter cell water/min for fully in- duced cells) were determined for *E. coli* CA8000. $V_{\text{max}}^{\text{efflux}}$ was assumed to equal $V_{\text{max}}^{\text{influx}}$, and K_m^{efflux} was assumed to equal 84 mM. For purposes of calculation, it

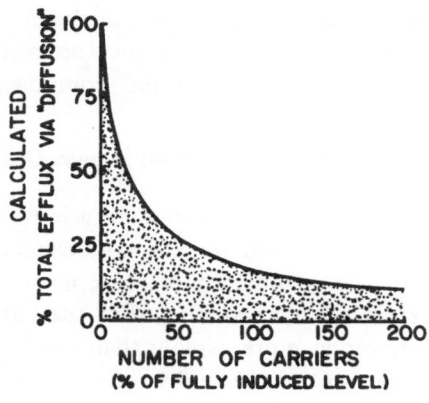

was assumed that the external substrate concentration (S_o) was 1 mM. For additional information, see Maloney and Wilson (1973). From Maloney and Wilson (1973).

to the number of carriers. Thus for an equivalent transport system, one with a low V_{max} compared to the diffusion pathway, one might propose a mechanism in which the carrier cannot participate in efflux. This would be incorrect. Other kinds of problems are encountered when the "diffusion" pathway makes only a small contribution to efflux. In this case, conditions which change the number of carriers or alter the V_{max} of the system in other ways do not result in similar changes in the steady-state level of accumula- tion. For example, *E. coli* with two copies of the *lac* operon does not show a 2-fold increase in the capacity to accumulate thiomethylgalactoside. In fact, after evaluating the parameters of the steady-state equation, Maloney and Wilson (1973) concluded that such partial diploids would be expected to show only a 20% increase in their ability to concentrate thiomethyl- galactoside.

It should also be appreciated that the importance of the "diffusion" pathway will vary with the particular substrate employed. It was mentioned previously that thiomethylgalactoside has a larger "diffusion" pathway than does lactose. This accounts for the observation (Maloney and Wilson, 1973) that cells with only 3% of the normal number of lactose membrane carriers accumulate lactose to about 40% of the normal level but can accumulate thiomethylgalactoside to only about 10–15% of the normal level.

This discussion should make it apparent that maintenance of the steady state during active transport represents the combined effects of specific, carrier-mediated substrate fluxes and the movement of substrate through the "diffusion" pathway. For this reason, conclusions should not be drawn solely from steady-state measurements if one has not evaluated the contribution of the "diffusion" component for transport.

2.5. Demonstration of Exchange Diffusion

The term "exchange diffusion" (Levi and Ussing, 1948) refers to the observation that the entry or exit of substrate may be stimulated by the presence of substrate on the opposite face of the membrane. This has been explained by assuming that the carrier–substrate complex traverses the membrane more rapidly than the free carrier. Thus the rate-limiting step in entry or exit would be the return of the empty carrier, and this may be accelerated if the carrier returns combined with a substrate molecule (Jacquez, 1961). This phenomenon is well described for transport systems in animal cells (Heinz, 1954; Levine *et al.*, 1965), and there is evidence showing that exchange diffusion also occurs for the lactose transport system in *E. coli* (Robbie and Wilson, 1969).

Experimentally, the demonstration of exchange diffusion in bacteria depends on eliminating the recapture event. The interplay between exchange diffusion and recapture is illustrated by the following two examples. In studying the effects of internal substrate on the influx step, one preloads with nonradioactive material and then exposes cells to labeled substrate. A stimulation of the entry of labeled substrate would result from exchange diffusion. However, an inhibition of entry of labeled material would occur because of the recapture of *un*labeled molecules in the periplasmic space. The net result might be that entry rates in control and preloaded cells would be the same. Very different results would be found in efflux studies. In efflux experiments, the presence of unlabeled substrate in the external medium stimulates exit by blocking recapture of labeled molecules. Superimposed on this would be an additional enhancement of efflux due to exchange diffusion. Thus, because of the presence of recapture, one might observe little exchange diffusion for entry but very strong exchange diffusion for exit. In fact, these are the observations which have been made under the usual conditions for studying the lactose transport system (Kepes, 1960; Koch, 1963; Winkler and Wilson, 1966; Robbie and Wilson, 1969).

Robbie and Wilson (1969) were able to separate the effects of exchange diffusion and recapture in their study of entry via the lactose transport system. They used cells preequilibrated with a high concentration of unlabeled substrate (the external concentration would be about 20 times the effective K_m for entry) and followed the initial rate of entry of labeled

Table I. Effect of Preloading on Initial Rates of Entry[a]

External substrate concentration (mM)	Ratio preloaded entry rate/nonpreloaded entry rate
0.075	1.16 ± 0.07
0.20	1.46 ± 0.06
1.0	1.38 ± 0.03
5.0	1.73 ± 0.04
20.0	1.84 ± 0.07

[a] For the indicated external substrate concentrations, the rates of entry of [^{24}C]thiomethyl-galactoside into cells of *E. coli* ML 308-831 were determined. See caption of Fig. 1 for details of the preparation of cell suspensions. Cells either were preloaded with 20 mM nonradioactive substrate or were used without preloading. The initial rate of entry was calculated from the linear increase in the intracellular concentration of radioactive substrate found during the first 30 s of incubation. Results were expressed as the ratio of the initial rate of entry into preloaded cells over that determined for nonpreloaded cells. Each value given is the mean ± SE for nine observations, each of which is the ratio of preloaded over non-preloaded entry rates measured on the same day with the same batch of cells. From Robbie and Wilson (1969).

substrate. Under these conditions, competition for influx by nonradioactive molecules being recaptured is effectively eliminated since the external concentration of substrate is high compared with the K_m for entry. Using this procedure, it could be shown that exchange diffusion results in about a 2-fold stimulation of the initial rate of entry of substrate (Table I). This technique allows the demonstration of exchange diffusion with respect to the entry step. As yet, it has not been possible to study exchange diffusion for exit because of the recapture process. Presumably, one might approach this problem by using protoplasts or cells in which the cell wall is abnormally "leaky." One would expect to find that exchange diffusion for exit also has about a 2-fold effect on rates of efflux.

According to the interpretation of exchange diffusion effects, the phenomenon occurs because of the different mobilities of the carrier and carrier–substrate complex. It may be noted, therefore, that exchange diffusion may vary according to whether the carrier–substrate complex has a mobility greater than, equal to, or less than that of the free carrier. Thus some substrates may show "positive" exchange diffusion, as in the examples given above, others may show no exchange diffusion, and still others may give "negative" exchange diffusion. Examples of each of these categories have been found for the substrates of the lactose transport system (Robbie and Wilson, 1969). Finally, it should be pointed out that the occurrence of exchange diffusion in the lactose transport system does not affect the form of the steady-state equation given in the previous section. These transmembrane effects may alter rates of substrate movement but maintenance of the final steady-state gradient is independent of their presence since they must act equally on both entry and exit.

2.6. Demonstration of Counterflow

The term "counterflow" refers to a special case of competitive inhibition. In counterflow experiments a transient accumulation against a concentration gradient may be demonstrated in a system which carries out facilitated diffusion. This phenomenon was first predicted by Widdas (1952) from theoretical considerations and later experimentally demonstrated for the sugar transport system in the erythrocyte (Park et al., 1956). In bacteria, counterflow has been demonstrated for the lactose transport system of E. coli (Koch, 1963; Winkler and Wilson, 1966; Wong and Wilson, 1970; Kepes, 1971) and for the galactose transport system of S. lactis (Kashket and Wilson, 1972a).

The experimental observation is that if cells are preloaded with a high internal concentration of (unlabeled) substrate A, centrifuged, and then resuspended in a low concentration of (labeled) substrate B, there is a transient accumulation of B within cells. This is readily explained by a competition for efflux between preloaded molecules of substrate A, at high internal concentration, and newly entering molecules of substrate B, which are at a low internal concentration. The appropriate control in such experiments is to examine the behavior of cells which are not preloaded. In these cells, substrate B would not accumulate but would reach an internal concentration equal to that in the external medium due to simple facilitated diffusion.

Wong and Wilson (1970) have carefully examined counterflow via the lactose transport system in metabolically poisoned E. coli [see also Wilbrandt's (1972) thorough analysis of counterflow in the erythrocyte]. Incubation of E. coli K12 and ML strains with 30 mM azide for 30 min at room temperature completely blocks their capacity to accumulate substrate, yet other criteria, including counterflow experiments, indicate that functional membrane carriers are present. Other starvation procedures, such as that devised by Koch (1971), block both accumulation and carrier activity; counterflow cannot be demonstrated in such cells (Olden and Wilson, unpublished results). Counterflow in azide-treated cells is demonstrated by the following method. After azide treatment, cells are incubated with unlabeled substrate (at a concentration about 20 times the K_m for exit, which is the same as the K_m for entrance under these conditions) for a time sufficient to allow equilibration of substrate across the membrane (usually 30 min at room temperature). Cells are then centrifuged into a compact pellet. High centrifugation speeds (40,000g for 10 min) are required to give a small cell pellet with as little as possible trapped extracellular fluid. After centrifugation, the supernatant is carefully decanted and all residual extracellular fluid is wiped from the sides of the tube. Cells are then quickly resuspended in medium containing azide and labeled substrate (at a concentration about equal to the K_m of the system) and samples are rapidly removed for filtration and washing. The results of such an experiment are given in Fig. 3, which describes the counterflow of thiomethylgalactoside. In control cells, not preloaded, internal substrate equilibrated with the external medium (0.5 mM thiomethylgalactoside). In cells preloaded with 20 mM unlabeled thiomethylgalactoside, there was a rapid and transient accumulation of labeled material. The maximal concentration gradient was about 10-fold and was attained within the first 30 s of incubation.

The major factor determining the shape of the counterflow curve appears to be the rate of efflux of preloaded substrate. Thus when the

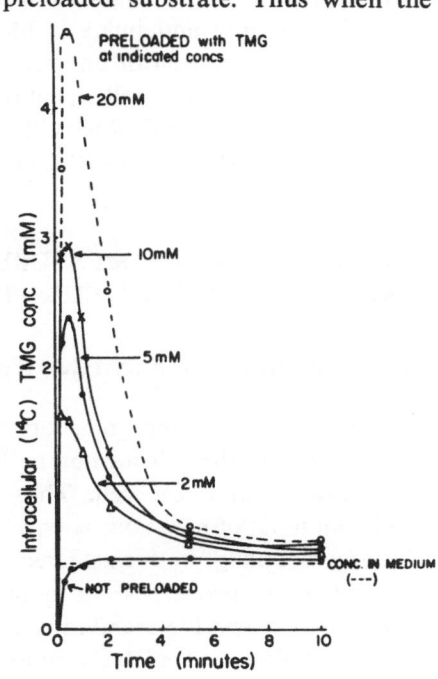

Fig. 3. Counterflow of thiomethylgalactoside. Cell suspensions of *E. coli* ML 308-831 were prepared as described in Fig. 1 except that glucose was not present. Washed cells were exposed to 30 mM azide in the presence of 0, 2, 5, 10, or 20 mM thiomethylgalactoside (TMG) for 30 min at 25°C. Following centrifugation (see text), each pellet was resuspended in medium containing 30 mM azide plus 0.5 mM [^{14}C]thiomethylgalactoside at 25°C. At the indicated times, samples containing 200 μg dry weight of cells were filtered and washed. From Wong and Wilson (1970).

number of carriers is reduced, lowering the rate of efflux of unlabeled material, the time course of the counterflow curve is considerably extended and the peak internal concentration of labeled substrate is somewhat lowered. Since efflux is determined by both carrier-mediated and "diffusion" components, efflux of preloaded material may also be reduced if that substrate has a slow "diffusion" rate. Thus counterflow of labeled thiomethylgalactoside is dramatically stimulated when cells are preloaded with lactose rather than with thiomethylgalactoside itself, presumably because efflux of lactose via "diffusion" is much slower than for thiomethylgalactoside. From these kinds of considerations it follows that counterflow will be difficult to demonstrate in cases where the "diffusion" component for efflux is large compared to the carrier-mediated pathway.

In bacteria, a quantitative analysis of the kinetics of counterflow is difficult because of the presence of recapture. Nevertheless, the demonstration of counterflow constitutes a powerful test for the presence of functional membrane carriers. This conclusion may not be easily established solely from the kinetics of substrate movement in a facilitated diffusion system.

In the method outlined above, transient accumulation of substrate occurs because influx proceeds while efflux is temporarily blocked. One may also demonstrate counterflow in the reverse direction, when efflux proceeds normally and influx is blocked. In this case one exposes cells to a high concentration of unlabeled substrate after labeled substrate has equilibrated across the cell membrane. After the addition of unlabeled material, there is a transient net efflux of labeled substrate which eventually reenters the cell (Wong and Wilson, 1970).

3. TECHNIQUES FOR MEASURING TRANSPORT WITHOUT ACCUMULATION OF SUBSTRATE

3.1. Nonmetabolizable Substrates: Facilitated Diffusion

Facilitated diffusion transport systems mediate the equilibration of substrate across the cell membrane but do not catalyze the accumulation of substrate within the cell. Thus facilitated diffusion systems do not require an input of metabolic energy. However, aside from the phenomenon of energy coupling, both facilitated diffusion and active transport systems have the same properties. In both instances, substrate movements are mediated by a specific membrane carrier which shows the characteristics expected of biological catalysts: substrate specificity, saturation kinetics, etc.

In bacteria, only the glycerol transport system of *E. coli* is known to be a facilitated diffusion system in normal cells (Sano *et al.*, 1968). Under certain conditions, however, a number of *active* transport systems in bacteria will carry out facilitated diffusion. In metabolically poisoned *E. coli*, the lactose transport system mediates the facilitated diffusion, but not the accumulation, of its substrates. Similarly, in *S. lactis* the galactose transport system behaves as a facilitated diffusion mechanism if cells are not given a fermentable carbon source. In each of these systems, the accumulation of substrate appears to be driven by a protonmotive force (Mitchell, 1963; West and Mitchell, 1973; West and Wilson, 1973; Kashket and Wilson, 1973, 1974). Thus one may anticipate that when other proton-linked active transport systems are described they, too, will behave as facilitated diffusion systems in the absence of energy coupling. For this reason, this section deals primarily with methods used to characterize systems, such as the lactose transport system, which normally carry out active transport but which behave as facilitated diffusion systems when separated from the energy coupling mechanism.

There are two major technical problems in demonstrating facilitated diffusion in such systems. One is that the bacterial cell water represents only a small fraction of the total water present in the sample taken for filtration. Thus an adequate washing protocol becomes essential. The sample must be placed in the center of the presoaked filter so that no fluid is trapped beneath the chimney, and, to avoid resuspension of the cells, the wash should not begin until the sample has been completely filtered. To increase the sample size, one may increase the cell density in the assay mixture so that the sample contains about 1 mg dry weight of cells, if a larger pore size (1.2 μm) filter is used. The larger pore size filter permits rapid filtration and washing with larger sample sizes but still retains 95% of the cells in *E. coli*.

The second major major problem stems from the fact that the K_m for efflux is the same as the K_m for influx in a facilitated diffusion system. Thus substantial efflux may occur even at early times when the internal concentration of substrate is low. To overcome this problem, one might block the exit of labeled substrate by using cells preloaded with a high concentration of unlabeled substrate, as discussed above (Section 2.3.1); this, of course, runs the "risk" of exchange diffusion. In addition, one might use the technique described by Winkler and Wilson (1966), which allows a sample to be taken as early as 5–6 s after the addition of labeled material. This is done as follows. Cells are preloaded with nonradioactive substrate (at a concentration about 20 times the K_m of the system) at 23°C.

External substrate is removed by centrifugation at 0°C followed by one wash with ice-cold medium and preloaded cells are finally resuspended in a small volume of iced medium. The efflux of preloaded substrate is greatly reduced at this low temperature. Using a chilled microliter syringe (0°C), at 0 time 50 µl of the cell suspension is added to 0.45 ml of assay medium containing radioactive substrate at the desired assay temperature. Then as soon as possible (5–6 s) most of the sample is taken up in a pasteur pipette (equilibrated to the assay temperature) and placed on the center of a pre-wetted filter for filtration and washing (also at the assay temperature). The actual volume delivered to the filter can be calculated from the difference between the counts in the initial sample and the counts remaining in the test tube and pasteur pipette. This method was successfully used in the measurement of both the K_m and the V_{max} of the lactose transport system in energy-poisoned cells (Winkler and Wilson, 1966). The dramatic effect of preloading is illustrated by the data given in Table II. In the absence of preloading, the V_{max} for entry in poisoned cells was less than 10% of that found in normal cells. In preloaded cells, however, the V_{max} for facilitated diffusion was equal to the V_{max} measured for the initial rate of entry into cells which could accumulate substrate. It should be noted that preloading had no apparent effect on the K_m values for entry.

Table II. Kinetic Parameters of the Lactose Membrane Carrier in Actively Transporting and Metabolically Poisoned Cells[a]

	K_m (mM)	V_{max} (mmol/liter cell water/min)
Actively transporting cells		
Nonpreloaded	0.6 ± 0.25	53 ± 17
Metabolically poisoned cells		
Nonpreloaded	0.9 ± 0.17	4.7 ± 1.8
Preloaded	1.0 ± 0.23	51 ± 16

[a] Cells of *E. coli* ML 308-225 (constitutive for the lactose transport system but lacking β-galactosidase) were grown to exponential phase in medium 63 containing 1% of a casein hydrolysate. After harvesting, cells were washed and resuspended in medium 63 containing 0.05 mg/ml chloramphenicol. The initial rates of entry of [14C]lactose were measured at external substrate concentrations from 0.1 to 2 mM. Rates of entry into metabolically poisoned cells (30 mM azide plus 1 mM iodoacetate for 30 min at 23°C) were determined either directly or after cells had been preloaded with 25 mM nonradioactive lactose. Actively transporting cells were sampled between 5 and 15 s; poisoned cells were sampled between 5 and 10 s. See text for details of sampling and preloading methods. The K_m and V_{max} values given are expressed as the mean ± SE of from three to five determinations. From Winkler and Wilson (1966).

Fig. 4. Facilitated entry of thiomethylgalactoside. Experimental details were as described in Fig. 3 except that the poisoned cells were not loaded with nonradioactive substrate. The *E. coli* strains used were ML 308 (constitutive for the lactose transport system) and its transport-negative derivative ML 35. From Wong and Wilson (1970).

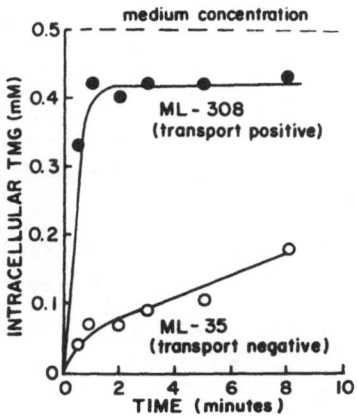

For some purposes it is not necessary to obtain such accurate measurements of the kinetic properties of the facilitated diffusion system. In such cases, three other tests serve to demonstrate that the entry of substrate is mediated by a specific membrane carrier. In mutants (or uninduced cells) lacking the membrane carrier, the rate of entry of substrate is substantially reduced. It was this kind of test that was necessary before the entry of glycerol into *E. coli* could be attributed to a facilitated diffusion system (Sano *et al.*, 1968). Figure 4 shows thiomethylgalactoside entry into metabolically poisoned wild-type cells of *E. coli* and into cells lacking the lactose transport system. In the wild-type cells, substrate equilibrated across the membrane within several minutes, whereas in the mutant final equilibration of substrate required a much longer time, about 20 min in this experiment (not shown). In a second test, one demonstrates that rates of entry of a substrate are reduced in the presence of chemically related compounds. For example (Fig. 5), in *S. lactis* deprived of an energy source, internal thiomethylgalactoside equilibrates with the medium concentration within 2 min and in the presence of competitive inhibitors this equilibration is markedly delayed. Finally, one should be able to demonstrate the phenomenon of counterflow in a transport system which mediates facilitated diffusion (Fig. 3).

3.2. Metabolizable Substrates

In studying transport systems, one attempts to choose conditions so that measurement of substrate fluxes reflects the activity of the transport system itself rather than other cellular processes. In general, this is done in one of two ways. One approach is to study transport under conditions

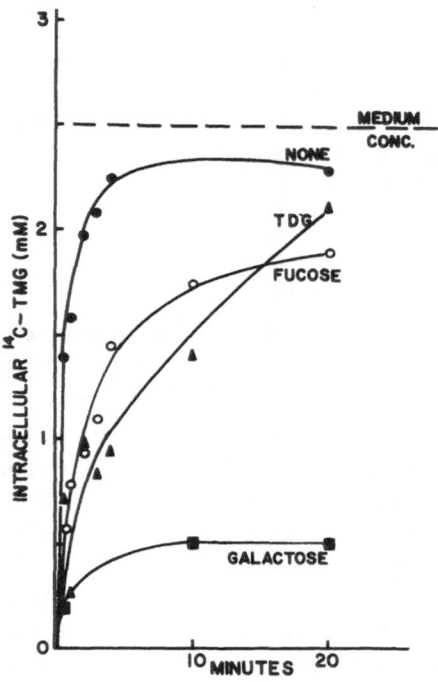

Fig. 5. Effect of various sugars on thio-methylgalactoside equilibration in *S. lactis*. Cells of *S. lactis* were grown to the exponential phase as described by Kashket and Wilson (1972a). Washed cells were incubated in 0.1 M sodium phosphate, pH 7, containing 1 mM magnesium chloride and either 20 mM D-fucose, 20 mM thiodigalactoside (TDG), 10 mM galactose, or no further additions. The reaction was initiated by adding [^{14}C]thiomethylgalactoside (TMG) and 0.1-ml samples containing 260 μg/dry weight of cells were removed at the indicated times for filtration and washing. From Kashket and Wilson (1972a).

where the substrate cannot be chemically altered by the cell, for example, by choosing nonmetabolizable analogues of the physiological substrate or by studying transport in mutant strains which lack the enzyme(s) responsible for utilization of the natural substrate. Alternatively, one may choose conditions under which the substrate is rapidly converted to some derivative within the cell. In this case, the rate-limiting step in incorporation would be the transport step and initial rates of entry would represent the activity of the membrane carrier even though subsequent metabolism occurs reflecting nontransport events. Intermediate situations, in which rates of transport are closely matched by metabolic capacities, give results which are often difficult to interpret, since variations in the rates of entry and internal pool sizes may represent changes in either the transport or utilization of substrate.

One major advantage in studying the lactose transport system in *E. coli* is that one may designate experimental conditions so that the substrate is nonmetabolizable *or* subject to rapid utilization. Thiogalactoside analogues of lactose are not hydrolyzed by the internal β-galactosidase and accumulate within the cell without chemical modification. However, other substrates of the transport system, such as lactose and o-nitrophenyl-galactoside (ONPG), are rapidly hydrolyzed within cells containing β-galactosidase. Use of these latter substrates allows one to study the translocation step apart from the overall process of accumulation (Herzenberg, 1959; Koch, 1963; Maloney and Wilson, 1974).

The principle behind the assay of ONPG entry is as follows. ONPG enters the cell via the lactose transport system and is hydrolyzed within the cell by β-galactosidase, releasing free galactose and o-nitrophenol. o-Nitrophenol is yellow and easily measured spectrophotometrically by its absorption at 420 nm. Since cells with no permeability barrier hydrolyze ONPG at least 10 times more rapidly than intact cells, the rate-limiting step in o-nitrophenol release by intact cells is the entry step rather than the hydrolytic step. Thus the rate of appearance of o-nitrophenol is the same as the rate of entry of ONPG.

Conditions for measuring ONPG entry are similar to those for assaying accumulation by this transport system. Washed cells are resuspended in medium 63 with chloramphenicol and maintained on ice until use. The assay mixture consists of medium 63 (pH 7) containing cells, ONPG (1–2 mM final concentration), chloramphenicol (50 μg/ml), and β-mercaptoethanol (about 1 mM final concentration). The reaction may be initiated by adding either cells or substrate. The rate of appearance of o-nitrophenol can be followed continuously in a recording spectrophotometer or by removing aliquots at various times and mixing with 3 vol of 0.6 M sodium carbonate (or other base) to terminate the reaction. After carbonate addition, cell debris may be eliminated by centrifugation or filtration, or appropriate blanks can be prepared. In this assay it is important to evaluate the nonspecific hydrolysis of ONPG which is due to the presence of a small fraction of external β-galactosidase and the presence of "diffusion" pathways for ONPG entry. This nonspecific component of ONPG hydrolysis is found by assaying parallel tubes to which thiodigalactoside (5 mM final concentration) has been added in addition to ONPG. This galactoside has an affinity for the lactose membrane carrier (Rickenberg et al., 1956; Kepes, 1960) and inhibits transport of ONPG but does not inhibit hydrolysis of ONPG by β-galactosidase (Rickenberg et al., 1956). Such blank values are typically 5–10% of the carrier-mediated component of ONPG hydrolysis when assays are done within 1–2 h of harvesting, but may be higher with older preparations.

Lactose entry may be assayed by a procedure analogous to that for ONPG entry. The products of intracellular lactose hydrolysis are free galactose and glucose. Under suitable conditions, the rate of lactose entry may be followed by the measurement of the appearance of either of these hexoses. In many instances, the release of galactose may be determined, since this sugar is not utilized by most strains of *E. coli* unless cells were pregrown in the presence of inducers of the *gal* operon. For technical reasons it is more convenient to assay the release of glucose. In this case, one must use a mutant strain such as strain GN2 which lacks the capacity to metabolize glucose (Fraenkel *et al.*, 1964). Strain GN2 lacks both enzyme I of the phosphotransferase system (Tanaka *et al.*, 1967) and glucokinase (Fraenkel *et al.*, 1964) and therefore cannot phosphorylate this sugar. The protocol for assaying lactose entry is as follows. The reaction is initiated by placing washed cells into medium 63 (pH 7) containing 50 mM NaCl, 50 μg/ml chloramphenicol, and 10 mM lactose. To terminate the reaction, *p*-chloromercuribenzoate is added to a final concentration of 0.1 mM. At this point, cells may be removed by filtration or by precipitation with an excess of barium hydroxide followed by neutralization with zinc sulfate. The concentration of free galactose is determined using galactose dehydrogenase as described by Wallenfels and Kurz (1962). The concentration of glucose is readily determined using glucose oxidase (the Glucostat method: Worthington Biochemicals). Under these conditions, the rate-limiting step in the appearance of free hexose is the transport step since cells with no permeability barrier hydrolyze lactose 3–5 times more rapidly than do intact cells. This is shown by assaying lactose hydrolysis after cells have been exposed to toluene and deoxycholate (Novick and Weiner, 1957). The assay of lactose entry has proved useful in demonstrating that the activity of the lactose membrane carrier may be affected by metabolic events within the cell. The mechanism responsible for this control is not yet completely understood but it appears that the rate of translocation of substrate is linked to the activity of some component of oxidative metabolism (Maloney and Wilson, 1974).

In *E. coli*, β-glucosides are transported by a group translocation mechanism, the phosphotransferase system (Kundig and Roseman, 1971), and appear within the cell as phosphorylated derivatives. In an elegant approach to the study of the translocation of substrates by this system, Fox and Wilson (1968) developed an assay analogous to the ONPG entry assay described above. They found that *p*-nitrophenyl-β-glucoside was a substrate for this transport system and that the phosphorylated form was hydrolyzed by internal phospho-β-glucosidase, liberating free *p*-nitrophenol and glucose-6-phosphate. In cell extracts, phospho-β-glucosidase activity was 5-fold higher than that required to account for the hydrolysis of *p*-nitrophenyl-β-glucoside by intact cells. Thus, for intact cells, the rate of appearance of

p-nitrophenol could be taken as a measure of the rate of entry (and phosphorylation at the membrane) of substrate.

4. ION MOVEMENTS ASSOCIATED WITH SOLUTE ACCUMULATION

Exploration of the relationships between ion movements and nonelectrolyte transport is proving valuable in elucidating the mechanisms by which metabolic energy is coupled to active transport in bacteria. There is now good evidence to support the suggestion of Mitchell (1963) that bacteria couple the movement of protons down their electrochemical gradient with the uphill transport of various substrates such as sugars and amino acids. The cotransport of protons and β-galactosides was first shown in *E. coli* (West, 1970) and later in *S. lactis* (Kashket and Wilson, 1973). In *E. coli*, one proton is translocated for every galactoside molecule (West and Mitchell, 1972, 1973). A detailed description of the techniques for measuring the translocation of protons across cell membranes is beyond the scope of this review. One useful approach was introduced by Mitchell and Moyle (1967), who incubated mitochondria in lightly buffered solutions and measured the pH of the medium under a variety of conditions. Their method was then extended to bacteria by Scholes and Mitchell (1970a) and has been used with *E. coli* by West (1970), West and Mitchell (1972, 1973), and West and Wilson (1973), and with membrane vesicles from *E. coli* by Hertzberg and Hinkle (1974) using the apparatus described by Thayer and Hinkle (1973).

The electrochemical force that drives proton-coupled substrate accumulation in bacteria, termed the "protonmotive force" (ΔP), is equal to the sum of the electrical potential ($\Delta \psi$) across the membrane and the chemical potential represented by the transmembrane pH gradient (Mitchell, 1966). This relationship can be expressed as

$$\Delta P = \Delta \psi - Z \Delta \, \mathrm{pH}$$

where ΔP and $\Delta \psi$ are expressed in millivolts and Δ pH is equal to the inside pH minus the outside pH. Z is a factor equal to 2.303 RT/nF, where R is the gas constant, T is the absolute temperature, F is Faraday's constant, and n is equal to 1. Z has a value of about 59 at 25°C.

By measuring both the hydrogen ion gradient and the electrical potential across the cell membrane of *S. lactis*, with due awareness of the limitations inherent in the various methods, the protonmotive force has been shown to equal the chemical potential represented by the accumulation of a nonmetabolizable sugar (Kashket and Wilson, 1973, 1974). In principle, this approach may be extended to other bacteria and to other transport systems. For this reason, the following sections present a discussion of the methods currently available for the measurement of both $\Delta\psi$, the electrical potential across the bacterial cell membrane, and Δ pH, the transmembrane pH gradient.

4.1. Membrane pH Gradient

4.1.1. Dye Methods

Based on a method devised for mitochondria (Chance and Mela, 1966, 1967), Scholes and Mitchell (1970b) have used the indicator dye bromocresol purple as a qualitative index of the internal pH of bacteria. A change in the internal pH was indicated by a change in absorbance at a wavelength appropriate for the dye, using a dual wavelength spectrophotometer. The external pH was monitored simultaneously, allowing an estimate of the behavior of the membrane pH gradient under a variety of experimental conditions.

In yeast, Kotÿk (1963) has used the dye bromothymol blue to quantitatively measure the internal pH of cells. The method is based on the assumption that the membrane is freely permeable to the acid (undissociated) form of the dye but impermeable to the charged species. This means that the concentration of the uncharged form will be the same on both sides of the membrane whereas the charged species will distribute itself in direct proportion to the membrane pH gradient. Thus an acidic dye will accumulate in cells when the internal pH is more alkaline than the external pH. If the pK_a' of the dye is sufficiently lower than the external or internal pH, then the concentration of the ionized form on either side of the membrane will, for practical purposes, equal the total concentration of the dye on that side of the membrane. In Kotÿk's experiments, yeast was incubated with a low concentration of bromothymol blue (about 1 μM, $pK_a' = 4.0$), and samples were filtered through Millipore filters (0.45 μm pore size) at intervals. After alkalinization, the concentration of the dye remaining in the filtrate was determined spectrophotometrically at 592 nm. The internal pH (pH_i) was then calculated from the relationship

$$pH_i = pH_o + \log\left[\frac{C_i}{C_o}(1 + 10^{pK_a'-pH_o}) - 10^{pK_a'-pH_o}\right]$$

where pH_o is the extracellular pH (monitored), pK'_a is the negative logarithm of the apparent ionization constant of the dye, C_o is the external concentration of the dye, and C_i is the internal concentration of the dye, calculated from the initial C_o minus the value of C_o at the indicated time. A correction was made for dye absorbed by cell components, as measured with sonically disrupted cell suspensions. The intracellular water volume was taken from previously determined values (63–71% of the total cell volume, depending on the composition of the external medium).

4.1.2. DMO Method

The method most frequently used to estimate the intracellular pH of bacteria is the determination of the distribution of the weak acid 5,5-dimethyl-2,4-oxazolidinedione (DMO). The undissociated forms of a number of weak acids and bases, including DMO, readily equilibrate across cell membranes, while their ionized forms apparently are impermeable (Caldwell, 1956). Since the degree of ionization of DMO on each side of the cell membrane depends on the pH, the concentration of DMO (ionized plus un-ionized forms) differs on the two sides if the pHs are different. In most experiments, extracellular water volume greatly exceeds intracellular water volume, so that, in practical terms, the DMO method is useful only when the cell's interior is more alkaline than the exterior, leading to the accumulation of this weak acid.

Waddell and Butler (1959) first used the DMO method to determine the internal pH of skeletal muscle. The method has since been extended to a variety of animal cells (reviewed by Waddell and Bates, 1969). The values obtained with the DMO method have been found to agree well with those given by other methods, including direct measurements using microelectrodes. Previous methods for estimating the intracellular pH of animal cells have included the measurement of the distribution of bicarbonate and of ammonium ions, direct measurement of the pH of tissue extracts, microinjection of pH indicators, and the use of microelectrodes (Caldwell, 1956).

The various properties of DMO necessary for its use as an indicator of the membrane pH gradient in bacteria have been carefully tested by Harold et al. (1970). Using Streptococcus faecalis, these workers showed that DMO is metabolically inert at concentrations less than 1 mM, that cells do not absorb DMO nonspecifically, and that DMO is not actively transported. The permeability of the cell membrane to the undissociated acid, but not to the ionized form, was indicated by the fact that DMO stabilized osmotically sensitive protoplasts at an alkaline pH but not at an acid pH. Harold *et al.* (1970) were able to use the DMO method to show that, during the fermentation of glucose, cells of *S. faecalis* maintained a membrane pH gradient of 0.5–1.0 pH unit (inside alkaline), whereas the fermentation of arginine did not yield a pH gradient. Cells of *S. lactis* have been found to behave similarly (Kashket and Wilson, 1974).

Measurement of the distribution of [^{14}C]DMO between cells and medium is complicated by the presence of extracellular fluid that contaminates the separated cell mass. Since the cells cannot be washed after filtration, it is necessary to monitor the trapped extracellular fluid with nonmetabolizable, impermeable ^{3}H-labeled molecules which can occupy all the external space up to the plasma membrane. A suitable compound for experiments with streptococci is D-sorbitol. Freshly chromatographed [^{3}H]inulin can also be used, giving slightly lower values for extracellular water space in *S. lactis*, presumably because inulin can penetrate only to the cell wall and not to the plasma membrane (Kashket and Wilson, unpublished results).

A typical experiment measuring the DMO distribution between cells and medium is given in Table III. Cells of *S. lactis* were incubated in a buffered medium containing [^{14}C]DMO and D-[^{3}H]sorbitol. At the desired time, cells were separated from the medium by rapid filtration through *dry* membrane filters, about 1 μm pore size, but *not washed* (the distribution of DMO is quite rapid, occurring within 1 min, and washing would cause a loss of DMO from within the cell). The filters with the cells and contaminating medium were then counted in a scintillation counter for both ^{3}H and ^{14}C. The ^{3}H content of the filtered cells indicates the trapped extracellular fluid and must be determined for each sample. Since the proportion of intracellular water in the reaction mixture is very small (less than 0.5% of the total fluid volume), the specific activities of the labeled compounds in the extracellular fluid are taken from the ^{3}H and ^{14}C content of aliquots of the original cell suspension. The internal pH (pH_i) is then calculated from the equation (Waddell and Butler, 1959)

$$pH_i = pK_a' + \log\left\{\left[\frac{C_t}{C_e}\left(1 + \frac{V_e}{V_i}\right) - \frac{V_e}{V_i}\right][10^{pH_o - pK_a'} + 1] - 1\right\}$$

where pH_o is the extracellular pH, pK_a' is the negative logarithm of the apparent ionization constant of DMO, approximately 6.32 at 25°C and 6.21 at 37°C (Addanki *et al.*, 1968), V_e and V_i are, respectively, the extracellular and intracellular fluid volumes of the filtered cells, C_t is the con-

Table III. Determination of Transmembrane pH Gradient by the DMO Method[a]

	Experiment A		Experiment B	
	^{14}C	^3H	^{14}C	^3H
	Average corrected cpm			
0.1 ml cells + medium (cpm/100 μl)	37,748	24,929	35,605	25,099
0.2 ml filtered cells (cpm/sample)	3,469	1,867	4,053	2,692

From D-[^3H]sorbitol cpm

V_e = extracellular water (μl)	7.49	10.73
V_i = intracellular water (μl)	0.77	0.77
$V_e + V_i$ = total water (μl)	8.26	11.49
V_e/V_i	9.75	13.97

From [^{14}C]DMO cpm

C_t = DMO concentration in $(V_e + V_i)$ (cpm/μl)	420	352
C_e = DMO concentration in V_e (cpm/μl)	377	356
C_t/C_e	1.11	0.99
$10^{pH_o - pK'_a} + 1$	1.06	11.48
$y = \left[\dfrac{C_t}{C_e} \left(1 + \dfrac{V_e}{V_i} \right) \right] - \dfrac{V_e}{V_i} [10^{pH_o - pK'_a} + 1] - 1$	1.35	8.55
$\log y$	0.13	0.93
$pH_i = \log y + pK'_a$ of DMO	6.45	7.25
pH_o	5.11	7.34
$\Delta pH = pH_i - pH_o$ (alkaline inside)	1.34	−0.09

[a] Cells of *S. lactis* 7962 (200 ml) were grown to exponential phase as described by Kashket and Wilson (1973). After harvesting, they were washed with 30 ml of a solution (pH 7.0) containing 150 mM sodium chloride and 2 mM glycylglycine and resuspended in 5 ml of the same medium as a concentrated stock. The suspension was incubated at 23°C for automatic titration to pH 7.0 (Radiometer pH Stat) with 0.1 N sodium hydroxide until acid production had ceased (about 50 min). Aliquots were then withdrawn and placed into tubes containing 5 ml 0.1 M sodium phosphate at pH 5.2 (experiment A) or pH 7.4 (experiment B). In addition, each tube contained [^{14}C]DMO (8 μCi/μmol, 50 μM final concentration) and D-[^3H]sorbitol (5 μCi/μmol, 100 μM final concentration). After 3–5 min, four aliquots of 0.2 ml (containing 530 μg dry weight of cells, equivalent to 0.77 μl cell water) were filtered through dry Millipore filters (1.2 μm pore size) without washing. The filters were placed into scintillation vials with 0.1 ml of water to equalize quenching. Two 0.1-ml samples of the cell suspension were also taken for direct counting (cells plus medium). After sampling, the pH of the cell suspension was determined (pH$_o$). Bray's solution was used for counting ^3H and ^{14}C radiation. The scintillation counter was set so that no ^3H radiation was found in the ^{14}C channel. The counts given in the table have been corrected for ^{14}C "spillover" into the ^3H channel (about 9% of the total ^{14}C counts).

centration of DMO in the total sample water, and C_e is the concentration of DMO in the medium.

4.1.3. Methylamine Method

While the internal pH of bacteria is usually the same or more alkaline than that of the external medium, there are circumstances in which the inside becomes more acid than the outside. Such is the case, for example, when valinomycin is added to cells of *S. lactis* suspended in low K^+ media (Kashket and Wilson, 1973). The efflux of potassium ions down their concentration gradient establishes a membrane potential (interior negative) which causes net proton uptake. The interior of the cell becomes more acid than the exterior by about 1.5 pH units (Table IV).

To measure a membrane pH gradient (inside acid) it is necessary to determine the accumulation of a base. The methylamine method has been adapted from experiments with chloroplasts (Rottenberg *et al.*, 1972). Two other methods, of possible use in bacterial systems, have also been used for the continuous measurement of the membrane pH gradient (inside acid) in chloroplasts. One method involves measuring the uptake of fluorescent amines with high pK_a' values (Schuldiner *et al.*, 1972). The other makes use of cationic electrodes to follow the disappearance of NH_4^+ from the medium (Rottenberg and Grunwald, 1972); as these authors note, the development of ammonium-selective electrodes should obviate the use of Na^+- and K^+-free media that is necessary when using cationic electrodes to measure ammonium ions.

As with the distribution of weak organic acids, the distribution of the base methylamine reflects the transmembrane pH gradient, assuming equilibration of the unprotonated (uncharged) form of the amine across the cell membrane. The pK_a' of methylamine (10.6) is much higher than the pH values usually encountered. Thus, both inside and outside the cell, the uncharged form will be at very low concentration compared to that of the charged species. In practice, this means that the ratio IN/OUT of total methylamine will very nearly equal the ratio IN/OUT of the hydrogen ion.

An illustration of the use of methylamine as an indicator of the transmembrane pH gradient (inside acid) is given in Table IV. The experiment consists of incubating cells of *S. lactis* in buffer containing [^{14}C]methylamine and D-[^{3}H]sorbitol, the latter to monitor the extracellular fluid trapped on the membrane filter. After separation of the cells from the medium by filtration, the filter is counted for both ^{3}H and ^{14}C as in the DMO method.

Table IV. Determination of the Transmembrane pH Gradient by the Methylamine Method[a]

	Before valinomycin		20 min after valinomycin	
	^{14}C	^{3}H	^{14}C	^{3}H
	corrected cpm			
0.1 ml cells + medium	7,258	20,008	7,258	20,008
0.1 ml filtered cells	646	1,375	1,736	1,505
Intracellular water (μl)	0.57		0.57	
Extracellular water (μl)	6.87		7.52	
Extracellular methylamine (cpm)	499		546	
Intracellular methylamine (cpm)	147		1190	
[Methylamine]$_{in}$ (cpm/μl)	259		2100	
[Methylamine]$_{in}$/[methylamine]$_{out}$	3.57		28.9	
ΔpH (acid inside)	0.56		1.46	

[a] Exponential-phase cells of *S. lactis* were washed with 0.1 M tris-HCl, pH 8.0, resuspended as a concentrated stock in 0.1 M tris-HCl, pH 7.1, and incubated for 1 h at 23°C. To begin the experiment, cells were placed in 0.1 M tris-HCl, pH 7.15, together with 3 μM [^{14}C] methylamine (17 μCi/μmol) and 100 μM D-[^{3}H]sorbitol (5 μCi/μmol). Aliquots of 0.1 ml (containing 390 μg dry weight of cells, equivalent to 0.56 μl cell water) were filtered through dry Millipore filters (1.2 μm pore size) without washing and the filters were counted for radioactivity as described in Table III. After 10 min, 10 μl of 10 mM valinomycin in 95% ethanol was added and further samples taken after an additional 20 min. Samples of 0.1 ml of the cell suspension were also taken for direct counting (cells plus medium). All manipulations were done at 23°C. The value for Δ pH (inside acid) was taken to equal log ([methylamine]$_{in}$/[methylamine]$_{out}$).

In this example, proton uptake by cells is the result of a membrane potential, negative inside, effected by efflux of the potassium ion in the presence of valinomycin. It should be noted that the concentration of methylamine must be kept low as it may act as an uncoupler; in chloroplasts, concentrations greater than 0.1 mM methylamine uncouple photophosphorylation (Rottenberg *et al.*, 1972).

4.2. Membrane Potential

Nutrients such as sugars and amino acids have been shown to accumulate within bacterial cells and bacterial membrane vesicles in response to a membrane potential, negative inside (Kashket and Wilson, 1972*b*; Niven *et al.*, 1973; Ashgar *et al.*, 1973; Hirata *et al.*, 1974; Hinds and

Brodie, 1974; Niven and Hamilton, 1974). As outlined below, there are a number of techniques which enable one to estimate the size of such an electrical potential across the bacterial membrane. These include measurements of (1) the accumulation of artificial permeant cations, (2) the distribution of naturally occurring cations (e.g., K^+) across membranes rendered permeable to them, and (3) fluorescence changes of dyes that respond to the membrane potential.

Early work on bacterial membrane potentials was reviewed by Cirillo (1966). Later work was summarized by Harold (1972) in his interpretation of energy transductions by bacterial membranes in terms of the chemiosmotic hypothesis.

4.2.1. Permeant Cations

The distribution of permeant cations between cells and the medium can be interpreted in terms of a membrane potential by means of the Nernst equation:

$$\Delta\psi = - \frac{RT}{zF} \ln \frac{[\text{cation}]_{in}}{[\text{cation}]_{out}}$$

where $\Delta\psi$ is the membrane potential (in millivolts), R is the gas constant, F is Faraday's constant, T is the absolute temperature, and z is the valency of the cation. (For simplicity, the concentrations rather than the activities of the cations are considered here.) For monovalent ions at 25°C, the Nernst equation is more simply expressed as

$$\Delta\psi = -59 \log \frac{[\text{cation}]_{in}}{[\text{cation}]_{out}}$$

Following the pioneering approach of Russian workers investigating the membrane potential of mitochondria (see reviews by Liberman and Skulachev, 1970; Skulachev, 1971), Harold and Papineau (1972) first used the lipid-soluble cation dibenzyldimethylammonium (DDA^+) to study the membrane potential of bacteria. Because of the complex ring structure of DDA^+, its positive charge is delocalized and the ion can pass through biological membranes. Since DDA^+ is not metabolized, its partition between the intracellular and extracellular phases should monitor the electrical potential across the cell membrane. Thus DDA^+ will accumulate within cells possessing a membrane potential, inside negative.

Harold and Papineau (1972) determined the accumulation of DDA^+ within cells of S. faecalis by measuring the disappearance of DDA^+ from the medium. This was done by following the change in absorbance at 262 and 266 nm of cell filtrates after interfering materials (such as nucleotides) had been removed by an anion exchange resin. A similar technique was used to study the distribution of another permeant cation, triphenylmethyl-phosphonium ($TPMP^+$), whose concentration in cell filtrates could be estimated by absorbancy measurements at 266.5 and 270.5 nm. During glucose fermentation, DDA^+ was shown to diffuse passively into K^+-depleted, Na^+-loaded cells of *S. faecalis*, reaching a concentration ratio IN/OUT of about 420. Assuming that the distribution of DDA^+ was in equilibrium with a Nernst potential, this indicated a membrane potential of 155 mV (inside negative). The rate, but not the extent, of DDA^+ uptake was increased by the addition of small amounts of the anion tetraphenyl-boron (TPB^-). The permeant cation $TPMP^+$ was equally accumulated by cells but did not require the presence of catalytic amounts of TPB^- for rapid accumulation. Both DDA^+ and $TPMP^+$ showed an initial nonspecic binding to cells that was unaffected by TPB^-. In K^+-replete cells, DDA^+ was not accumulated unless valinomycin was added to render the membrane highly permeable to the potassium ion (see following section). During glucose fermentation in such valinomycin-treated cells (external pH of 7.5), the final DDA^+ concentration ratio indicated a membrane potential of about -165 mV, whereas the potassium distribution suggested a membrane potential of about -195 mV.

Tritium-labeled DDA^+ has been used by Hirata *et al.* (1973) to estimate the membrane potential of vesicles from *E. coli* prepared by the method of Kaback (1971). After the nonspecific binding of $[^3H]DDA^+$ to membrane filters was subtracted from the experimental values, it could be shown that in Na^+-loaded vesicles, with D-lactate present as an oxidizable energy source, $[^3H]DDA^+$ (TPB^- present) accumulated 60-fold, indicating a membrane potential of about -100 mV. In K^+-replete vesicles and in the absence of D-lactate, $[^3H]DDA^+$ was accumulated after the addition of valinomycin but, again, only in the presence of TPB^-; the $[^3H]DDA^+$ ratio IN/OUT was 10, equivalent to a potential of -60 mV. In each of the examples cited above, the accumulation of $[^3H]DDA^+$ was abolished by the addition of the proton conductor carbonylcyanide-*m*-chlorophenylhydrazone or tetra-chlorosalicylanilide.

As Harold and Papineau (1972) and Hirata *et al.* (1973) discuss in detail, the limitations of the permeant cation DDA^+ as a measure of the membrane potential are considerable. Among the uncertainties are (1) the apparent requirement for catalytic amounts of the anion TPB^-, (2) the difference in behavior of Na^+- vs. K^+-loaded vesicles or cells, (3) the considerable nonspecific absorption of DDA⁺ to membrane filters and cells, and (4) the discrepancies in the ratios IN/OUT of DDA⁺ and K⁺ in potassium-permeable cells. Until these problems are resolved, it may be best to use the distribution of permeant cations such as DDA⁺ and TPMP⁺ as a qualitative index of the membrane potential, rather than as a quantitative tool.

4.2.2. Potassium Distribution Method

The concentration ratio IN/OUT of K^+ may also be used as a measure of the membrane potential, if the cell membrane is sufficiently permeable to the potassium ion. To make bacterial membranes highly permeable to this cation, one usually employs the potassium ionophore valinomycin (Harold, 1970) at a final concentration of 1–10 μM. Special procedures are not required for the use of valinomycin with bacteria such as the streptococci (Harold and Baarda, 1967; Kashket and Wilson, 1972b). Lawford and Haddock (1973) report similar behavior by K12 cells of *E. coli*. However, West and Mitchell (1972), using ML *E. coli*, found that cells had to be pretreated according to the method of Leive (1968) before valinomycin was effective.

Measurement of the potassium distribution ratio is most easily done using flame photometry. An example of this is given in Table V. Cells are separated from the medium by filtration but *not* washed. To liberate internal potassium, the filters are placed in glass tubes containing 1 or 2 drops of *n*-butanol and the tubes incubated in a boiling water bath for about 15 min. After cooling, a known volume of LiCl is added, and the tubes are vigorously mixed to remove all material from the filter. Automatic diluter attachments on clinical flame photometers cannot be used since the absolute amount of K^+ available for analysis is too small. Therefore, it is necessary to include LiCl (to a final concentration equivalent to that reached by the diluter attachment) in all assay solutions including standards. If the external K^+ concentration is very low compared to that found internally, the contribution of K^+ made by extracellular fluid trapped on the filter may be small enough to be disregarded. However, in other cases, as in the example given in Table III, one must account for such contamination. This may be done by including labeled D-sorbitol or inulin in the reaction mixture and measuring the radioactivity in the LiCl resuspension.

One source of error in estimating the membrane potential from the potassium distribution ratio is the possibility that the concentration of potassium in the bulk medium is lower than in the fluid immediately

Table V. Determination of Membrane Potential Using Potassium Distribution Method, and Correlation of the Membrane Potential with Fluorescence Changes of the Dye CC_6[a]

Vessel No.	$[K^+]_{out}$ (mM)	Minutes after valinomycin	$K^+/1.0$ ml filtered cells (μmol)		$[K^+]_{in}$ (mM)	$\dfrac{[K^+]_{in}}{[K^+]_{out}}$	$\Delta\psi$ (mV)	Fluorescence[b] (units)
			Total	Extracellular				
A	11.54	1	0.210	0.075	333	28.9	85.6	12.8
		2	0.198	0.075	302	26.2	83.6	12.9
		3	0.238	0.117	298	25.8	83.3	12.5
		4	0.238	0.124	280	24.2	81.6	12.3
B	22.35	1	0.335	0.219	287	12.8	65.3	10.4
		2	0.325	0.220	259	11.6	62.8	10.2
		3	0.325	0.235	223	10.0	58.9	9.3
		4	0.325	0.245	198	8.9	55.9	9.3
C	34.94	2	0.350	0.270	197	5.6	44.3	8.4
		3	0.305	0.216	220	6.3	47.1	7.9
		4	0.435	0.304	233	6.7	48.8	7.6

[a] Cells of *S. lactis* (200 ml) were grown for 11 h to early stationary phase using methods previously described (Kashket and Wilson, 1973). After harvesting, cells were washed with 0.1 M sodium phosphate, pH 7.0, and resuspended in 10 ml of this same buffer. One-tenth milliliter of saturated (about 5 M) glucose and 0.1 ml of 1 M potassium chloride were added and the mixture was incubated for 30 min at 23°C. This incubation raised the internal potassium level from about 200 mM, typical of stationary-phase cells, to about 400 mM. The cells were then centrifuged, washed, and resuspended in 0.1 M sodium phosphate, pH 7.0. A stock suspension was prepared by adding 20 μl of 3.5 mM CC_6 in 95% ethanol to 4 ml of cells at a density of about 8.7 mg dry weight/ml. Aliquots of this stock suspension were diluted 30-fold for the determination of the potassium ratios IN/OUT (this table) and for assay with the fluorescent probe (Figs. 6 and 7). All manipulations, except for the initial centrifugations at 4°C, were carried out at 23°C. Each reaction vessel (A, B, and C) contained 0.3 ml stock suspension of cells plus CC_6, 0.9 ml of 1 M sodium chloride, potassium chloride to give the indicated medium potassium concentrations, 0.1 ml of 2 mM D-[³H]sorbitol (4 μCi/μmol), and 0.1 M sodium phosphate buffer, pH 7.0, to a final volume of 9 ml. After 10 min, valinomycin (15 μl of a 10 mM solution in 95% ethanol) was added and four 1-ml samples were removed at timed intervals. Cells were separated from the medium by Millipore filtration (1.2 μm pore size) without washing. To prepare cell extracts, the filters were placed in glass tubes and 1 drop of *n*-butanol was added to each filter. After incubation in a boiling water bath for 15 min, 2 ml of lithium chloride (30 ppm) was added. To prepare samples of the extracellular fluids, part of each reaction mixture was centrifuged and aliquots of the supernatants were made up to 30 ppm lithium chloride. The concentration of potassium in cell extracts and supernatants was determined with a flame photometer after appropriate dilution. Potassium ratios IN/OUT were calculated after correction for the potassium content of contaminating extracellular fluid, whose volume had been determined in each sample from the content of D-[³H]sorbitol. Cell density in the reaction mixture was equivalent to 0.41 μl cell water/ml. The membrane potentials given were calculated from the Nernst equation (see text). It should be noted that ionophores such as valinomycin or gramicidin adsorb to glass surfaces. They can be removed by rinsing with a polar solvent such as acetone or ethyl ether.

[b] See Figs. 6 and 7.

exterior to the plasma membrane in the periplasmic space. This would lead to an overestimation of the true value of the membrane potential, especially in experiments where cells are suspended in low-potassium media. It is likely that the higher the concentration of potassium in the medium, the smaller the difference between the concentration of potassium in the bulk medium and that in the periplasmic space. Another possible source of error is the difference between the activities of potassium ions inside and outside the cell. The activity of ions is affected by the total ionic strength of their environment and, unlike animal cells in isotonic media, bacterial protoplasm is usually hyperosmotic compared to many of the media employed in experiments (Mitchell and Moyle, 1956). In addition, if the principal anion within the cell is nucleic acid phosphate, the internal pH would affect the activity of the potassium ion. These possible errors tend to lead to an overestimation of the membrane potential, particularly at high K^+ ratios IN/OUT and in media of low osmolarity.

4.2.3. Fluorescent Dye Method

A number of recent observations, in both animal and bacterial cells, indicate the value of certain fluorescent dyes as tools to measure the electrical potential across cell membranes. Using human or *Amphiuma* red blood cells, Laris and Hoffman (1973) and Hoffman and Laris (1974) have studied the changes in the fluorescent intensity of the cyanine dye 1,1'-dihexyl-2,2'-oxacarbocyanine (CC_6) under conditions where the membrane potential could be manipulated experimentally. After the addition of valinomycin to a mixture of cells and dye, there was a decrease in fluorescence of CC_6 in media containing low potassium. The higher the potassium level in the medium, the smaller was the decrease in fluorescence. These red cells have a resting membrane potential (the Donnan potential) whose value may be estimated from the distribution of chloride, since these cells are known to be extremely permeable to anions. One important observation made by Hoffman and Laris (1974) was that at the external potassium concentration giving no fluorescence decrease (i.e., no change from the Donnan potential present before valinomycin addition) it could be shown that the membrane potential calculated from the potassium ratio IN/OUT was in agreement with that known from the chloride ratio OUT/IN in untreated cells. This was also true in the case of the *Amphiuma* giant red cell, where, in addition, the resting potential could be determined by direct measurements with microelectrodes (Hoffman and Laris, 1974). Sims *et al.* (1974) studied fluorescence changes of this dye and a number of related compounds in both red cells and artificial lipid vesicles treated with valino-mycin in media of different potassium levels. They found that such fluores-cence changes result primarily from a potential-dependent partition of the dye between cells and medium and postulated that the quenching of the cell-associated dye is due to the formation of dye aggregates with reduced fluorescence. As in the experiments with animal cells, when cells of *S. faecalis* or *S. lactis* were mixed with CC_6 the addition of valinomycin resulted in a decrease in fluorescence; the higher the level of potassium in the medium, the smaller was the quenching of fluorescence (Laris and Pershad-singh, 1974; Kashket and Wilson, 1974). By using this fluorescent probe, it has been possible to measure the membrane potential in bacteria under "physiological" conditions, such as during the fermentation of glucose or arginine by *S. lactis* (Kashket and Wilson, 1974).

The dye CC_6 is not entirely harmless to red blood cells; it causes increases in Na^+ and K^+ fluxes as well as some hemolysis (Hoffman and Laris, 1974). Some related dyes are less noxious (Sims *et al.*, 1974). In *S. lactis*, CC_6 at 15 μM appears to cause little damage to the cell membrane since glucose-energized sugar accumulation was decreased by only about 15% (Kashket and Wilson, 1974).

The following experiment illustrates the use of the fluorescent probe CC_6 in the determination of the membrane potential in *S. lactis*. In order to use the decrease in fluorescence of CC_6 as a quantitative measure of the membrane potential, fluorescence changes were correlated with the mem-brane potentials calculated from the distribution of potassium in valino-mycin-treated cells. As shown in Fig. 6 (see also Table V), when cells were mixed with CC_6 in sodium phosphate buffer, pH 7, and then treated with valinomycin, there was a decrease in fluorescence intensity. A maximum decrease of 25% was observed when no potassium was added to the medium (not shown). As the concentration of potassium in the medium was raised, there was a proportionately smaller decrease in fluorescence (curves A, B, and C). Also shown in Fig. 6 are the fluorescence changes observed in the absence of valinomycin when cells were energized by glucose fermenta-tion.

A calibration curve is given in Fig. 7. In this curve, the fluorescence change given on the ordinate refers to the difference between the fluorescence observed in the presence of valinomycin and the fluorescence observed after the addition of gramicidin. At this point, the protonmotive force was dissipated since this ionophore renders the membrane highly permeable to a variety of ions, including H^+, Na^+, and K^+ (Harold, 1970), and abol-ishes TMG accumulation. In the presence of gramicidin these ions should distribute themselves to equal ratios in/out, and quantitative measurements of sodium and potassium ratios after the addition of gramicidin confirmed this expectation. The ratios in/out of Na^+ or K^+ after gramicidin addition were unaffected by the addition of glucose. The change in fluorescence (increase) found after the addition of gramicidin was reproducible within one batch of cells and similar from one experiment to another. However, the change in fluorescence (decrease) observed after the addition of valino- mycin showed some variation, possibly because the initial membrane poten- tials varied. For this reason, all quantitative data refer to the fluorescence changes following the addition of gramicidin. In Fig. 7 the fluorescence

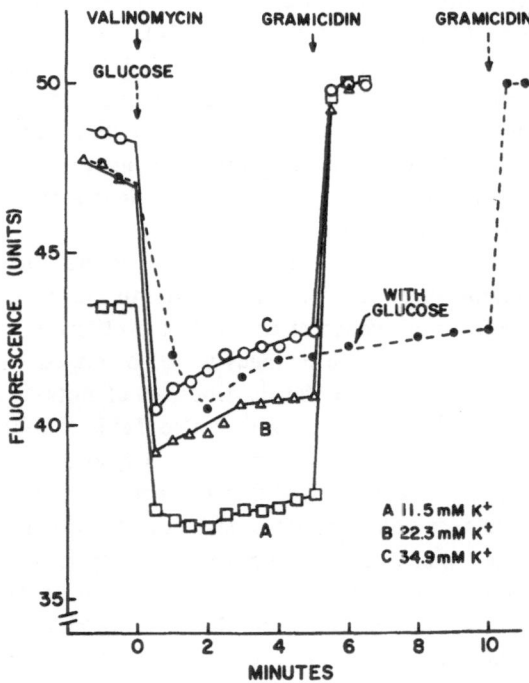

Fig. 6. **Fluorescence changes of CC_6**. Each cuvette contained 0.1 ml of the stock sus- pension of cells plus CC_6 (described in Table V), 0.3 ml 1 M sodium chloride, sufficient potassium chloride to give the indicated external potassium concentrations, and 0.1 M sodium phosphate, pH 7, to a final volume of 3 ml. After 10 min, valinomycin (5 μl of a 10 mM solution in 95% ethanol) or glucose (5 μl of about a 5 M solution) was added. This was designated 0 min. Gramicidin (5 μl of a 10 mM solution in 95% ethanol) was added when indicated. Fluorescence intensities were measured with an Aminco-Bowman spectrophotofluorometer at 450-nm excitation and 503-nm emission wavelengths. Fluores- cence readings were normalized to final values of 50 in arbitrary units.

Fig. 7. Relationship between the membrane potential and fluorescence change of CC_6. The
fluorescence changes (Δ fluorescence) given on
the ordinate are the values observed after
gramicidin addition minus the values observed
at timed intervals after valinomycin addition
for the experiment given in Fig. 6 (these fluorescence changes are also given in Table V). The
corresponding membrane potentials ($\Delta\psi$) were
calculated from the potassium ratios IN/OUT
measured in parallel reaction vessels (Table V).
The arrow indicates the fluorescence change
observed 6–10 min after the addition of glucose
(see Fig. 6) and corresponds to a membrane
potential of about -45 mV.

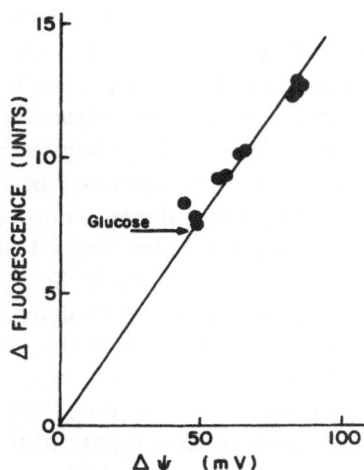

change found in glycolyzing cells is indicated by the arrow, and corresponds
to a membrane potential of about -45 mV (inside negative).

Similar patterns in the fluorescence of the dye CC_6 were observed
with exponential phase *S. lactis*, but after the initial decrease the fluorescence
rose again immediately, and within 15 min returned to approximately the
level seen before the addition of valinomycin. During this period, potassium
ions were shown to exit from the cells. One explanation for the difference
between early stationary-phase cells (Fig. 6) and the exponential-phase cells
is the older culture's relative impermeability to protons. Thus in early
stationary-phase cells the lack of a permeant counterion would prevent
the efflux of potassium, maintaining the membrane potential constant for
several minutes.

5. OSMOTIC SWELLING METHODS

5.1. Intact Cells

For over a hundred years the permeability of cells has been studied
by measuring the volume changes which result from exposure to solutions
of penetrating or nonpenetrating substances. In cells with a rigid cell wall,
such as plant and bacterial cells, the plasmolysis method has been extensively
utilized. This procedure involves placing the cell in a hypertonic solution
of the test substance. When a nonpenetrating material is used, water is
extracted osmotically from the cytoplasm, and the plasma membrane pulls

rapidly away from the cell wall, resulting in a new, smaller cell volume (plasmolysis). If a penetrating molecule is used, water is initially removed from the cell, but as the solute enters water reenters and the cytoplasmic volume returns to its original value (Fig. 8). In this case, the rate of return to the original volume becomes a measure of the rate of entry of the solute. These volume changes may be conveniently monitored by measuring transmitted light in a spectrophotometer. Swelling of the plasmolysed cell results in an increase in transmitted light, presumably due to a fall in refractive index and a reduction in light scattering. The volume change method has proved quite useful in studying membrane permeability in microorganisms, especially in cases where the rate of entry of solute is rapid. However, this technique can be used only in gram-negative bacteria. Gram-positive organisms do not readily show plasmolysis, probably because the plasma membrane adheres tightly to the rigid cell wall.

In 1903, Alfred Fisher demonstrated plasmolysis in gram-negative bacteria exposed to hypertonic solutions of sodium chloride or sucrose but not glycerol or urea. This was taken as evidence that such cells were very permeable to glycerol and urea but not to sodium chloride or sucrose. The experiments of Mitchell and Moyle (1956) give a quantitative illustration of such behavior (Figs. 9 and 10). Figure 9 shows the relative extinction observed when cells of *E. coli* were exposed to sodium chloride solutions of different osmotic strengths. As the osmotic strength of the medium was increased, plasmolysis resulted in an increase in the optical density of the suspension. In another experiment (Fig. 10), cells were exposed to 0.36 M erythritol. This concentration of erythritol yielded cells that were in a shrunken state initially. As erythritol entered the cell, causing its volume

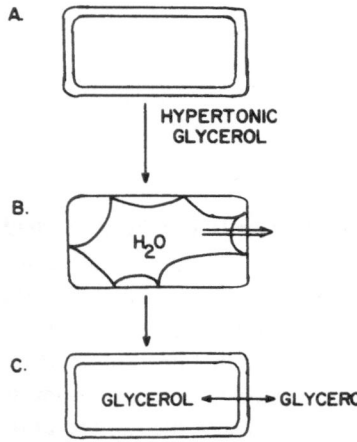

Fig. 8. A bacterium suddenly exposed to a hypertonic solution of glycerol. (A) Initially, the cell volume occupies all the space within the cell wall. (B) When hypertonic glycerol is added, water moves out of the cell, resulting in shrinkage. (C) As glycerol enters the cell, water reenters and the cytoplasm returns to its initial volume.

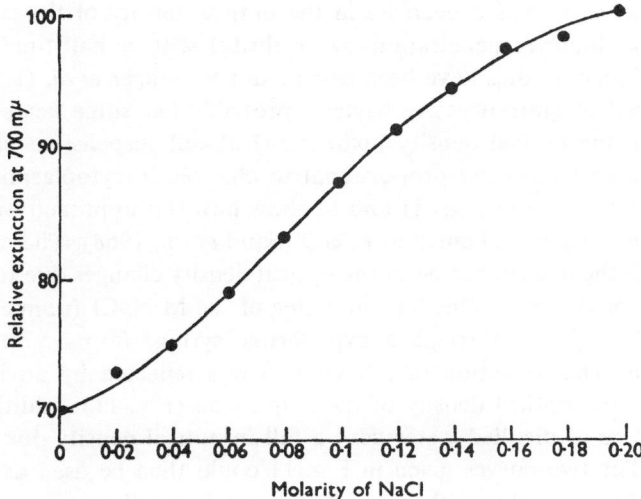

Fig. 9. Dependence of the extinction of suspensions of *E. coli* B at 700 nm on the sodium chloride concentration of the suspension medium. The suspension medium also contained 0.01 M phosphate at pH 6.8. From Mitchell and Moyle (1956).

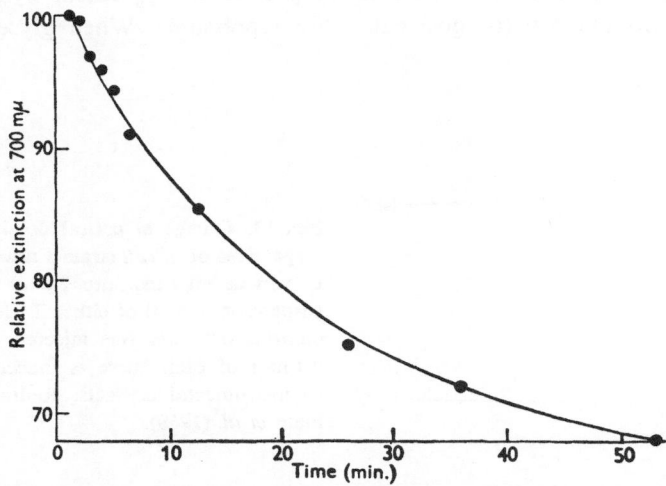

Fig. 10. Time course of the permeation of erythritol into *E. coli* B at 20°C measured by the change in extinction at 700 nm. At 0 time, cells were suspended in 0.36 mol/kg erythritol containing 0.01 M phosphate at pH 6.8. From Mitchell and Moyle (1956).

to increase, there was a decrease in the optical density of the suspension. These data indicate penetration by erythritol with a half-time of about 10 min. Similar studies have been carried out by Mager *et al.* (1956). Since the cell wall of gram-negative bacteria probably has some flexibility, such changes in the optical density (extinction) of cell suspensions should not be considered as directly proportional to changes in cytoplasmic volume.

The data given in Figs. 11 and 12 show how this approach was applied to the study of glycerol entry in *E. coli* (Sano *et al.*, 1968). The experiment began with the determination of the optical density changes due to maximal shrinkage of the cells. One-half milliliter of 2.4 M NaCl (nonpenetrating) was rapidly injected through a hypodermic syringe into 2.5 ml of a cell suspension. The reduction in cell volume was reflected by an immediate increase in the optical density of the suspension (Fig. 11); addition of the same volume of distilled water gave a fall in optical density due to simple dilution. The two curves given in Fig. 11 could then be used as points of reference for the volume changes occurring when cells were exposed to a concentrated solution of a penetrating substance. For example, if a slowly penetrating nonelectrolyte were added cells would rapidly shrink to the minimum volume (indicated by the NaCl curve) and then return to the initial volume (indicated by the H_2O curve). In such experiments, the rate of penetration of glycerol (added as 0.5 ml of a 4.8 M solution) was estimated under a variety of conditions. Cells with only a few membrane carriers for glycerol were obtained by growth in 1% casein hydrolysate plus 10 mM glucose (to give catabolite repression). When glycerol was

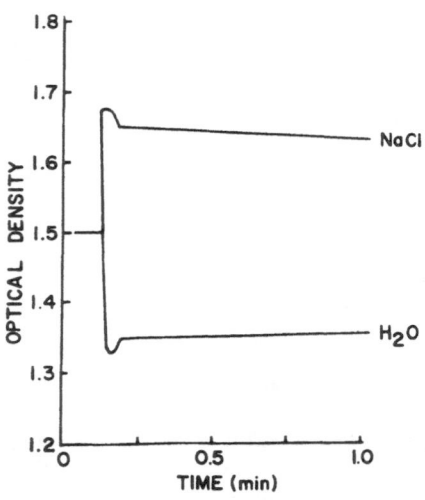

Fig. 11. **Change in optical density of cell suspensions of** *E. coli* **strain 1 after addition of various diluents.** Into 2.5 ml of a cell suspension, 0.5 ml of either 2.4 M sodium chloride or water was injected. The first 0.1 min of each curve is inaccurate due to instrumental artifacts. Redrawn from Sano *et al.* (1968).

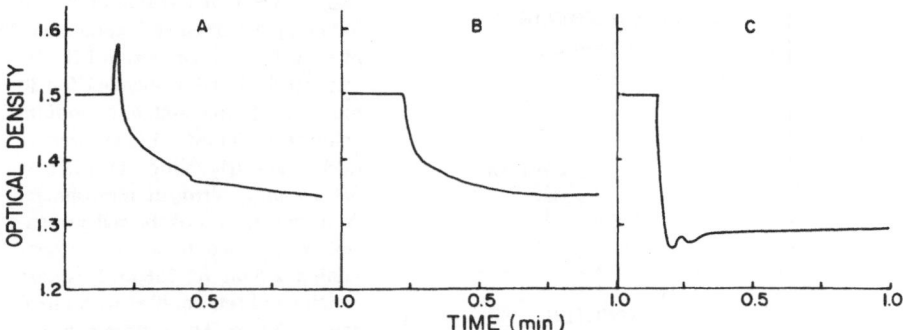

Fig. 12. Glycerol permeability of *E. coli* strain 1. Cells were grown using (A) 1% casein hydrolysate plus 10 mM glucose, (B) 1% casein hydrolysate, or (C) 20 mM glycerol. See text and Fig. 11 for additional information. Redrawn from Sano *et al.* (1968).

added to such cells, some shrinkage resulted, followed by a return to the initial volume (Fig. 12A). The half-time for return was about 15 s. When a similar experiment was done using fully induced (glycerol grown) cells, glycerol entered so rapidly that no shrinkage could be observed by this method (Fig. 12C). Equilibration of glycerol across the membrane had taken place within the 2–3 s required for mixing and for optical artifacts to disappear. Intermediate numbers of glycerol membrane carriers could be observed in cells grown in 1% casein hydrolysate without inducer and without glucose (Fig. 12B). From these kinds of experiments, it was concluded that *E. coli* possesses an inducible transport system for glycerol.

The study by West (1970) provides another example of the swelling of plasmolysed bacteria due to the penetration of a solute via a specific transport system. Intact cells of *E. coli* were placed in a solution of 0.386 M lactose. Swelling was observed when the transport system was functional but not when the membrane carriers were inactivated by formaldehyde (Fig. 13). Note that in this experiment no intracellular accumulation of lactose occurred since cells were energy-depleted by incubation under anaerobic conditions in the presence of iodoacetate.

5.2. Osmotically Sensitive Spheroplasts

Several microorganisms may be partially or completely freed from their rigid cell wall. Such cells are then osmotically sensitive and will swell or shrink in response to changes in external osmotic pressure. One of the best-studied examples of such osmotically fragile "protoplasts" is *Bacillus*

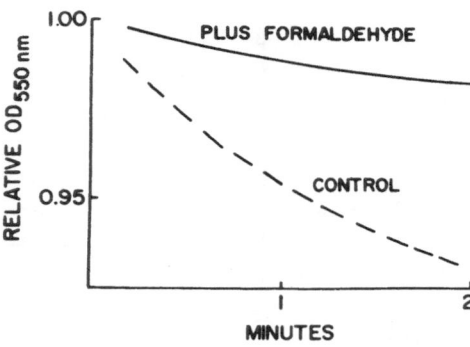

Fig. 13. Osmotic swelling of *E. coli* **following lactose entry.** Exponential-phase cells of *E. coli* strain ML 308 were washed and suspended for 30 min in a solution containing sodium chloride (150 mM), iodoacetate (1 mM), and tris buffer (1 mM) at pH 6.5 in a nitrogen atmosphere. At 0 time, 0.1 ml of the cell suspension was added to a cuvette containing 2.5 ml of 386 mM lactose which had been bubbled with nitrogen for 5 min. The lactose solution also contained 40 mM sodium thiocyanate and, where indicated, 30 mM formaldehyde. Redrawn from West (1970).

megaterium. Weibull (1953) showed that exposure of *B. megaterium* to lysozyme resulted in complete loss of the cell wall, leaving a spherical protoplast which could be lysed in solution of low osmotic pressure; such protoplasts lysed in 0.03 M phosphate buffer but were stabilized when 0.2 M sucrose was added.

Osmotically sensitive forms of *E. coli* may be prepared by treatment of cells with a combination of EDTA and lysozyme. The partial removal of the cell wall allows the cell to both swell and shrink. Such "spheroplasts" were used by Sistrom (1958) to demonstrate that lactose and thiomethylgalactoside accumulated by *E. coli* were osmotically active within the cell. In the presence of 0.01 M galactoside, swelling occurred in spheroplasts derived from induced cells, but not in spheroplasts which lacked the lactose membrane carrier. In addition, it could also be shown that the increase in water content of swollen spheroplasts corresponded to that expected if all the accumulated galactoside were in free solution within the cell.

6. REFERENCES

Addanki, S., Cahill, F. D., and Sotos, J. F., 1968, Determination of intramitochondrial pH and intramitochondrial-extramitochondrial pH gradient of isolated heart mitochondria by the use of 5,5-dimethyl-2,4-oxazolidinedione. 1. Changes during respiration and adenosine triphosphate-dependent transport of Ca++, Mg++ and Zn++, *J. Biol. Chem.* **243**:2337.

Ashgar, S. S., Levin, E., and Harold, F. M., 1973, Accumulation of neutral amino acids by *Streptococcus faecalis*: Energy coupling by a protonmotive force, *J. Biol. Chem.* **248**:5225.

Berger, E. A., 1973, Different mechanisms of energy coupling for the active transport of proline and glutamine in *Escherichia coli*, *Proc. Natl. Acad. Sci. USA* **70**:1514.

Britten, R. J., and McClure, F. T., 1962, The amino acid pool in *Escherichia coli*, *Bacteriol. Rev.* **26**:292.

Britten, R. J., Roberts, R. B., and French, E. F., 1955, Amino acid adsorption and protein synthesis in *Escherichia coli*, *Proc. Natl. Acad. Sci. USA* **41**:363.

Caldwell, P. C., 1956, Intracellular pH, *Int. Rev. Cytol.* **5**:229.

Carter, J. R., Fox, C. F., and Kennedy, E. P., 1968, Interaction of sugars with the membrane protein component of the lactose transport system of *Escherichia coli*, *Proc. Natl. Acad. Sci. USA* **60**:725.

Chance, B., and Mela, L., 1966, Hydrogen ion concentration changes in mitochondrial membranes, *J. Biol. Chem.* **241**:4588.

Chance, B., and Mela, L., 1967, Energy-linked changes of hydrogen ion concentration in submitochondrial particles, *J. Biol. Chem.* **242**:830.

Cirillo, V. P., 1966, Symposium on bioelectrochemistry of microorganisms. I. Membrane potentials and permeability, *Bacteriol. Rev.* **30**:68.

Cohen, G. N., and Rickenberg, H. V., 1956, Concentration spécifique réversible des amino acids chez *Escherichia coli*, *Ann. Inst. Pasteur* **91**:693.

Fisher, A., 1903, *Vorlesungen über Bakterien*, 2nd ed., Jena.

Fox, C. F., and Wilson, G., 1968, The role of a phosphoenolpyruvate-dependent kinase system in β-glucoside catabolism in *Escherichia coli*, *Proc. Natl. Acad. Sci. USA* **59**:988.

Fraenkel, D. G., Falcoz-Kelly, F., and Horecker, B. L., 1964, Utilization of glucose-6-phosphate by glucokinaseless and wild-type strains of *Escherichia coli*, *Proc. Natl. Acad. Sci. USA* **52**:1207.

Ganesan, A. K., and Rotman, R., 1965, Transport systems for galactose and galactosides in *Escherichia coli*. I. Genetic determination and regulation of the methyl-galactoside permease, *J. Mol. Biol.* **16**:42.

Harold, F. M., 1970, Antimicrobial agents and membrane function, in: *Advances in Microbial Physiology*, Vol. 4 (A. H. Rose and J. F. Wilkinson, eds.), pp. 45–104, Academic Press, New York.

Harold, F. M., 1972, Conservation and transformation of energy by bacterial membranes, *Bacteriol. Rev.* **36**:172.

Harold, F. M., and Baarda, J. R., 1967, Gramicidin, valinomycin and cation permeability in *Streptococcus faecalis*, *J. Bacteriol.* **94**:53.

Harold, F. M., and Papineau, D., 1972, Cation transport and electrogenesis by *Streptococcus faecalis*. I. The membrane potential, *J. Membr. Biol.* **8**:27.

Harold, F. M., Pavlasova, E., and Baarda, J. R., 1970, A transmembrane pH gradient in *Streptococcus faecalis*: Origin, and dissipation by proton conductors and N,N'-dicyclohexylcarbodiimide, *Biochim. Biophys. Acta* **196**:235.

Heinz, E., 1954, Kinetic studies on the "influx" of glycine-1-C^{14} into the Ehrlich mouse ascites carcinoma cell, *J. Biol. Chem.* **211**:781.

Hertzberg, E. L., and Hinkle, P. C., 1974, Oxidative phosphorylation and proton translocation in membrane vesicles from *Escherichia coli*, *Biochem. Biophys. Res. Commun.* **58**:178.

Herzenberg, L. A., 1959, Studies on the induction of β-galactosidase in a cryptic strain of *Escherichia coli*, *Biochim. Biophys. Acta* **31**:525.

Hinds, T. R., and Brodie, A. F., 1974, Relationship of a proton gradient to the active transport of proline with membrane vesicles from *Mycobacterium phlei, Proc. Natl. Acad. Sci. USA* **71**:1202.

Hirata, H., Altendorf, K., and Harold, F. M., 1973, Role of an electrical potential in the coupling of metabolic energy to active transport by membrane vesicles of *Escherichia coli, Proc. Natl. Acad. Sci. USA* **70**:1804.

Hirata, H., Altendorf, K., and Harold, F. M., 1974, Energy coupling in membrane vesicles of *Escherichia coli*. 1. Accumulation of metabolites in response to an electrical potential, *J. Biol. Chem.* **249**:2939.

Hoffman, J. F., and Laris, P. C., 1974, Determination of membrane potentials in human and *Amphiuma* red blood cells using a fluorescent probe, *J. Physiol. (London)* **239**:519.

Jacquez, J. A., 1961, Transport and exchange diffusion of L-tryptophan in Ehrlich cells, *Am. J. Physiol.* **200**:1063.

Kaback, R., 1971, Bacterial membranes, in: *Methods in Enzymology*, Vol. XXII (W. B. Jakoby, ed.), pp. 99–120, Academic Press, New York.

Kashket, E. R., and Wilson, T. H., 1972a, Role of metabolic energy in the transport of β-galactosides by *Streptococcus lactis, J. Bacteriol.* **109**:784.

Kashket, E. R., and Wilson, T. H., 1972b, Galactoside accumulation associated with ion movements in *Streptococcus lactis, Biochem. Biophys. Res. Commun.* **49**:615.

Kashket, E. R., and Wilson, T. H., 1973, Proton-coupled accumulation of galactoside in *Streptococcus lactis* 7962, *Proc. Natl. Acad. Sci. USA* **70**:2866.

Kashket, E. R., and Wilson, T. H., 1974, Protonmotive force in fermenting *Streptococcus lactis* 7962 in relation to sugar accumulation, *Biochem. Biophys. Res. Commun.* **59**:879.

Kepes, A., 1960, Études cinétiques sur la galactoside perméase d'*Escherichia coli, Biochim. Biophys. Acta* **40**:70.

Kepes, A., 1971, The β-galactoside permease of *Escherichia coli, J. Membr. Biol.* **4**:87.

Klein, W. L., and Boyer, P. D., 1972, Energization of active transport by *Escherichia coli, J. Biol. Chem.* **247**:7257.

Koch, A. L., 1963, The role of permease in transport, *Biochim. Biophys. Acta* **79**:177.

Koch, A. L., 1971, Energy expenditure is obligatory for the downhill transport of galactosides, *J. Mol. Biol.* **59**:447.

Kotȳk, A., 1963, Intracellular pH of baker's yeast, *Folia Microbiol.* **8**:27.

Kundig, W., and Roseman, S., 1971, Isolation of a phosphotransferase system from *Escherichia coli, J. Biol. Chem.* **246**:1393.

Laris, P. C., and Hoffman, J. F., 1973, Membrane potential in human red blood cells determined using a fluorescent probe, *Fed. Proc.* **32**:271 (abst.).

Laris, P. C., and Pershadsingh, H. A., 1974, Estimations of membrane potentials in *Streptococcus faecalis* by means of a fluorescent probe, *Biochem. Biophys. Res. Commun.* **57**:620.

Lawford, H. C., and Haddock, B. A., 1973, Respiration-driven proton translocation in *Escherichia coli, Biochem. J.* **136**:217.

Leder, I. G., 1972, Interrelated effects of cold shock and osmotic pressure on the permeability of the *Escherichia coli* membrane to permease accumulated substrates, *J. Bacteriol.* **111**:211.

Leive, L., 1968, Studies on the permeability change produced in coliform bacteria by ethylenediamine tetraacetate, *J. Biol. Chem.* **243**:2373.

Levi, H., and Ussing, H. H., 1948, The exchange of sodium and chloride ions across the fibre membrane of the isolated frog sartorius, *Acta Physiol. Scand.* **16**:232.

Levine, M., Oxender, D. L., and Stein, W. D., 1965, The substrate facilitated transport of the glucose carrier across the human erythrocyte membrane, *Biochim. Biophys. Acta* **109**:151.

Liberman, E. A., and Skulachev, V. P., 1970, Conversion of biomembrane-produced energy into electrical form. IV. General discussion, *Biochim. Biophys. Acta* **216**:30.

Mager, J., Kuczynski, M., Schatzberg, G., and Avi-dor, Y., 1956, Turbidity changes in bacterial suspensions in relation to osmotic pressure, *J. Gen. Microbiol.* **14**:69.

Maloney, P. C., and Wilson, T. H., 1973, Quantitative aspects of active transport by the lactose transport system of *Escherichia coli*, *Biochim. Biophys. Acta* **330**:196.

Maloney, P. C., and Wilson, T. H., 1974, Metabolic control of lactose entry in *Escherichia coli*, *Abst. Ann. Meeting, Am. Soc. Microbiol.*, item P 292.

Mitchell, P., 1963, Molecule, group and electron translocation through natural membranes, *Biochem. Soc. Symp.* **22**:142.

Mitchell, P., 1966, Chemiosmotic coupling in oxidative and photosynthetic phosphorylation, *Biol. Rev. Cambridge Philos. Soc.* **41**:445.

Mitchell, P., and Moyle, J., 1956, Osmotic function and structure in bacteria, in: *Bacterial Anatomy: Symposium of the Society for General Microbiology*, Vol. 6 (E. T. C. Spooner and B. A. D. Stocker, eds.), pp. 150–180, Cambridge University Press, Cambridge.

Mitchell, P., and Moyle, J., 1967, Acid-base titration across the membrane system of rat-liver mitochondria, *Biochem. J.* **105**:588.

Mitchell, P., and Moyle, J., 1968, Estimation of membrane potential and pH difference across the cristae membrane of rat liver mitochondria, *Eur. J. Biochem.* **4**:530.

Niven, D. F., and Hamilton, W. A., 1974, Mechanisms of energy coupling to the transport of amino acids by *Staphylococcus aureus*, *Eur. J. Biochem.* **44**:517.

Niven, D. F., Jeacock, R. E., and Hamilton, W. A., 1973, The membrane potential as the driving force for the accumulation of lysine by *Staphylococcus aureus*, *FEBS Lett.* **29**:248.

Novick, A., and Weiner, M., 1957, Enzyme induction as an all-or-none phenomenon, *Proc. Natl. Acad. Sci. USA* **43**:553.

Novotny, C. P., and Englesberg, E., 1966, The L-arabinase permease system in *Escherichia coli* B/r, *Biochim. Biophys. Acta* **117**:217.

Park, C. R., Post, R. L., Kalman, C. F., Wright, J. H., Johnson, L. H., and Morgan, H. E., 1956, The transport of glucose and other sugars across cell membranes and the effect of insulin, in: *Ciba Foundation Colloquia on Endocrinology: Internal Secretion of the Pancreas*, Vol. 9, pp. 240–260.

Rickenberg, H. V., Cohen, G. N., Buttin, G., and Monod, J., 1956, La galactoside permease d'*Escherichia coli*, *Ann. Inst. Pasteur* **91**:829.

Ring, K., 1965, The effect of low temperatures on permeability in *Streptomyces hydrogenans*, *Biochim. Biophys. Res. Commun.* **19**:576.

Robbie, J. P., and Wilson, T. H., 1969, Transmembrane effects of β-galactosides on thiomethyl-β-galactoside transport in *Escherichia coli*, *Biochim. Biophys. Acta* **173**:234.

Rotman, B., and Guzman, R., 1961, Transport of galactose from the inside to the outside of the cell, *Extrait de Pathologie-Biologie* **9**:806.

Rottenberg, H., and Grunwald, T., 1972, Determination of Δ pH in chloroplasts. 3. Ammonium uptake as a measure of Δ pH in chloroplasts and sub-chloroplast particles, *Eur. J. Biochem.* **25**:71.

Rottenberg, H., Grunwald, T., and Avron, M., 1972, Determination of Δ pH in chloroplasts. 1. Distribution of [^{14}C]-methylamine, *Eur. J. Biochem.* **25**:54.

Sano, Y., Wilson, T. H., and Lin, E. C. C., 1968, Control of permeation to glycerol in cells of *Escherichia coli*, *Biochem. Biophys. Res. Commun.* **32**:344.

Scholes, P., and Mitchell, P., 1970a, Acid-base titration across the plasma membrane of *Micrococcus denitrificans*: Factors affecting the effective proton conductance and the respiratory rate, *J. Bioenerg.* **1**:61.

Scholes, P., and Mitchell, P., 1970b, Respiration-driven proton translocation in *Micrococcus denitrificans*, *J. Bioenerg.* **1**:309.

Schuldiner, S., Rottenberg, H., and Avron, M., 1972, Determination of Δ pH in chloroplasts. 2. Fluorescent amines as a probe for the determination of Δ pH in chloroplasts, *Eur. J. Biochem.* **25**:64.

Schultz, S. G., and Curran, P. F., 1970, Coupled transport of sodium and organic solutes, *Physiol. Rev.* **50**:637.

Sims, P. J., Waggoner, A. L., Wang, C. H., and Hoffman, J. F., 1974, Studies on the mechanism by which cyanine dyes measure membrane potentials in red blood cells and phosphatidyl choline vesicles, *Biochemistry* **13**:3315.

Sistrom, W. R., 1958, On the physical state of intracellularly accumulated substrates of β-galactoside-permease in *Escherichia coli*, *Biochim. Biophys. Acta* **29**:579.

Skulachev, V. P., 1971, Energy transformations in the respiratory chain, *Curr. Top. Bioenerg.* **4**:127.

Sullivan, K. H., Jain, M. K., and Koch, A. L., 1974, Activation of the β-galactoside transport system in *Escherichia coli* by n-alkanols: Modification of lipid-protein interaction by a change in bilayer fluidity, *Biochim. Biophys. Acta* **352**:287.

Tanaka, S., Fraenkel, D. G., and Lin, E. C. C., 1967, The enzymatic lesion of strain MM-6, a pleiotropic carbohydrate-negative mutant of *Escherichia coli*, *Biochem. Biophys. Res. Commun.* **27**:63.

Thayer, W. S., and Hinkle, P. C., 1973, Stoichiometry of adenosine triphosphate-driven proton translocation in bovine heart submitochondrial particles, *J. Biol. Chem.* **248**:5395.

Waddell, W. J., and Bates, R. G., 1969, Intracellular pH, *Physiol. Rev.* **49**:285.

Waddell, W. J., and Butler, T. C., 1959, Calculation of intracellular pH from the distribution of 5,5-dimethyl-2,4-oxazolidinedione (DMO): Application to skeletal muscle of the dog, *J. Clin. Invest.* **38**:720.

Wallenfels, K., and Kurz, G., 1962, Über die Spezifität der Galaktosedehydrogenase aus *Pseudomonas saccharaphila* und deren Anwendung als analytisches Hilfmittel, *Biochem. Z.* **335**:559.

Weibull, C., 1953, The isolation of protoplasts from *Bacillus megaterium* by controlled treatment with lysozyme, *J. Bacteriol.* **66**:688.

West, I. C., 1970, Lactose transport coupled to proton movements in *Escherichia coli*, *Biochem. Biophys. Res. Commun.* **41**:655.

West, I. C., and Mitchell, P., 1972, Proton-coupled β-galactoside translocation in nonmetabolizing *Escherichia coli*, *J. Bioenerg.* **3**:445.

West, I. C., and Mitchell, P., 1973, Stoichiometry of lactose-H$^+$ symport across the plasma membrane of *Escherichia coli*, *Biochem. J.* **132**:587.

West, I. C., and Wilson, T. H., 1973, Galactoside transport dissociated from proton movement in mutants of *Escherichia coli*, *Biochem. Biophys. Res. Commun.* **50**:551.

Widdas, W. F., 1952, Inability of diffusion to account for placental glucose transfer in the sheep and consideration of the kinetics of a possible carrier transfer, *J. Physiol.* (*London*) **118**:23.

Wilbrandt, W., 1972, Coupling between simultaneous movements of carrier substrates, *J. Membr. Biol.* **10**:357.

Winkler, H. H., and Wilson, T. H., 1966, The role of energy coupling in the transport of β-galactosides by *Escherichia coli*, *J. Biol. Chem.* **241**:2200.

Wong, P. T. S., and Wilson, T. H., 1970, Counterflow of galactosides in *Escherichia coli*, *Biochim. Biophys. Acta* **196**:336.

Chapter 2

Preparation and Characterization of Isolated Intestinal Epithelial Cells and Their Use in Studying Intestinal Transport

GEORGE A. KIMMICH

Department of Radiation Biology and Biophysics
School of Medicine and Dentistry
University of Rochester
Rochester, New York

1. INTRODUCTION

Progress in our understanding of the nature of biological transport systems characteristic of intestinal tissue has been closely paralleled by the development of methods which allow various aspects of transport to be evaluated. Indeed, the rate of growth in our understanding of intestinal transport is to a high degree determined by the rate of development of innovative techniques coupled with ever-growing bodies of knowledge in biochemistry, physiology, and related biological disciplines. Any historical overview of a basic science will reflect the quantum leaps in understanding which are in part produced by innovations in methodology and are in part the driving force for further innovation. Necessity may be the mother of invention, but invention is both the mother and daughter of basic research.

As one might expect, the earliest attempts at probing intestinal transport phenomena were various *in vivo* techniques aimed at evaluating the kinds of biological molecules absorbed, their rate of transport, and the extent to which given concentrations of materials could be accumulated. The idea that biological tissue might survive and function after separation from the animal under study had not yet taken root, and, indeed, knowledge was not available even to allow the possibility. Consequently, various types of preparations involving study of intestinal segments in anesthetized animals or animals which had been surgically prepared beforehand formed the basis of the earliest investigations.

The first such experiments are commonly credited to Thiry and were reported in 1864. His procedure involved surgical preparation of a blind sac of intestine joined to the abdominal wall and open to the animal's exterior. Blood supply to the segment remained intact, and the loose ends of the intestine adjacent to the segment chosen for study were resected. Material could then be introduced to the permanent fistula, aliquots withdrawn at intervals, and net absorption determined by difference. The procedure was modified by Vella in 1888 so that both ends of the chosen segment were joined to the abdominal wall. Material could then be introduced at one end of the segment and collected at the other, and absorption again determined by difference. Cori (1925) introduced a method in which absorption could be studied from nonanesthetized animals which had not been surgically prepared beforehand. A measured volume of solution containing the solute for study was given by stomach tube. After an appropriate experimental interval, the animal was sacrificed and the amount unabsorbed determined by assay of the contents of the entire gastrointestinal tract. This natural perfusion procedure was an early forerunner of various *in vivo* perfusion techniques in which specific segments of intestine were perfused luminally with media of defined chemical composition (Sols and Ponz, 1947; Sheff and Smyth, 1955; Fullerton and Parsons, 1956; Jacobs and Luper, 1957). The perfusate was simply collected after a single pass or recycled, and assessments of absorption were determined from knowledge of initial and final solute concentrations and times and rates of perfusion. Data obtained from such perfusion studies in conjunction with those obtained from absorption of solutes introduced to tied loops of intestine in an anesthetized animal provided valuable evidence regarding the specificity and capacity of intestinal transport systems. Minor variations of these basic techniques made it possible to define the characteristics of absorption in different regions of the small intestine (Miller and Abbott, 1934; Abbott and Miller, 1936; Shay *et al.*, 1940; Cummins and Jussila, 1955).

Other methods allowed the possibility of learning the fate of absorbed solutes. Various techniques which involved cannulation of the portal vein (London, 1929) and the lymphatic system (Bollman *et al.*, 1948) were important in establishing the fact that absorbed sugars and amino acids reach the venous system draining the small intestine (Van Slyke and Meyer, 1912; von Mering, 1877; Dent and Schilling, 1949; Shoemaker *et al.*, 1963) while long-chain fatty acids, triglycerides, cholesterol, and other sterols are absorbed primarily via the lymphatic route (Bloom *et al.*, 1950; Bergstrom *et al.*, 1954; Blomstrand *et al.*, 1959; Biggs *et al.*, 1951; Chaikoff *et al.*, 1952). The general technique of cannulation was modified by Matthews and Smyth (1954) so that a mesenteric vein draining a specific section of intestine was cannulated and material absorbed from the ligated section could be collected quantitatively. The various cannulation techniques were important in ascertaining that most amino acids (Matthews and Smyth, 1954) and monosaccharide sugars (Shoemaker *et al.*, 1963; Kiyasu *et al.*, 1956) reach the bloodstream primarily in unaltered form. At the same time, they were important in ascertaining that oligosaccharides (Wilson and Vincent, 1955), oligopeptides (Levenson *et al.*, 1959; Wiggans and Johnston, 1959), and a portion of certain monosaccharides, such as fructose (Kiyasu and Chaikoff, 1957), appear in the portal blood as metabolites of the administered nutrient (see Table I). These observations

Table I. Metabolism of Various Solutes During Transfer from the Intestinal Lumen to Portal Blood or to the Serosal Compartment of *in Vitro* Preparations

Initial mucosal substrate	Products detected	Final concentration (mg/ml)	
		Mucosal	Serosal
5 mg/ml glycyl-leucine[a]	Glycine	1.1	1.75
	Glycyl-leucine	2.15	0.02
5.3 mg/ml glucose[b]	Glucose	3.3	6.4
	Fructose	—	—
6.8 mg/ml fructose[b]	Fructose	1.6	0.28
	Glucose	0.2	4.54
10 mg/ml sucrose[b]	Glucose	1.03	4.9
	Fructose	2.68	0.62
	Sucrose	—	—
10 mg/ml starch[b]	Starch	—	—
	Oligosaccharide	+	—
	Disaccharide	+	—
	Glucose	+++	++++

[a] Data from Newey *et al.* (1959).
[b] Data from Wilson and Vincent (1955).

led to the recognition that luminal disappearance of a solute did not necessarily define the characteristics simply of a transport system, but might also represent a degree of metabolic activity on the part of the intestinal epithelium superimposed on transport capability. Methods which would allow further insight into the integration between transport and metabolism were clearly needed.

The explosive development of modern biochemistry provided the appropriate environment for continued progress in probing intestinal transport phenomena. Particularly important was the growing recognition that tissue could be excised and maintained under laboratory conditions which permitted greater ease of manipulation and better definition of conditions during the interval of study than was possible with the *in vivo* techniques. A pioneering effort was reported in 1901 by Reid, who made use of a circular segment of intestinal tissue as a partition between two glass compartments and observed absorption of fluid against a hydrostatic pressure gradient. His apparatus was an early forerunner of modern Ussing-type flux chambers which are now commonly employed for monitoring transmural solute fluxes. A number of *in vitro* perfusion techniques were attempted in the first half of this century (*cf.* Ohnell, 1939), but they were difficult to manage and important questions were raised concerning the continued viability of the preparations. In 1949, Fisher and Parsons described a technique for circulating well-oxygenated saline medium on both sides of an intestinal loop maintained *in vitro* and were successful in demonstrating that the transport capability of the segment could be maintained. They observed transmural transfer of glucose against a concentration gradient and found that a very large portion of glucose disappearing from the lumen of the segment could not be accounted for in the serosal compartment. These were among the earliest observations which indicated that accumulation and/or metabolism of solute by the tissue represents a significant portion of the total solute disappearing from the lumen (see Table II). Several modifications of the Fisher and Parsons technique have been employed, notably those introduced by Wiseman (1953) and Darlington and Quastel (1953). All of these procedures have been valuable in demonstrating many of the characteristics of intestinal transport and the possibility of maintaining tissue in a viable transport state under *in vitro* conditions. The Quastel perfusion system was the first in which a dependence of transport on the presence of Na^+ was noted, an aspect which is now the focal point for investigations bearing on the molecular mechanism of transport.

Data obtained with the *in vitro* circulation techniques raised further questions. An active metabolic role for the tissue in supporting transport

Table II. Luminal Disappearance and Serosal Appearance of Sugar (mg/cm · h)
During Absorption by *in Vitro* Tissue Preparations

Sugar	Mucosal disappearance	Serosal appearance	Increase in tissue content	Calculated amount metabolized
Glucose[a]	2.13	0.98	0.16	0.99
Galactose[b]	1.20	0.53	0.20	0.47

[a] Data from Fisher and Parsons (1953a).
[b] Data from Fisher and Parsons (1953b).

had been clearly implicated by the fact that serosal to mucosal concentration gradients could be generated and that transmural transport could be inhibited by a variety of metabolic inhibitors which interfere with cellular energy-transducing capability. However, techniques were required which would allow sampling of the intestinal tissue itself if the nature of the transport system were to be further probed. Perfused tissue could, of course, be taken for assay purposes, but the technique allowed only a single tissue sample to be taken at one point in time following initiation of transport. Comparison between two different segments from the same animal was of questionable validity in light of Fisher and Parsons' (1949) evidence that transport and metabolic capability differ markedly in segments taken from different regions of the intestine.

Several *in vitro* techniques which allow tissue samples to be taken for assay of accumulated material appeared in the 1950s. The best known and most widely used is the everted sac technique of Wilson and Wiseman (1954). The procedure still requires one segment for each sample, but the segments are much smaller than those used for perfusion techniques and several sacs can be prepared from one intestinal region. A further modification was introduced by Agar *et al.* (1954) in which small transverse segments 1–2 mm wide were incubated with the appropriate radioactive substrate and tissue accumulation was determined at various intervals of time by rinsing the sections, extracting the solute absorbed, and quantitating the amount of isotope present. No closed serosal compartment was available in this system, making it the first preparation in which accumulation by the tissue itself rather than serosal transfer was the primary focus of study. Data obtained with these two procedures helped establish the fact that intestinal tissue itself accumulates solutes against a concentration gradient.

That the epithelial cells are primarily if not entirely responsible for tissue accumulation was suggested by the use of techniques in which epithelial cell sheets were separated from underlying tissue layers and the two preparations incubated separately (Bihler and Crane, 1962; Schultz *et al.*, 1966). Only the mucosal layer (stripped epithelial sheet) was capable of accumulating sugars and amino acids against a concentration gradient in a manner which depended on the presence of Na^+ and was inhibited by ouabain or various metabolic inhibitors. The submucosal layers showed no tendency to accumulate solutes regardless of the conditions employed. These facts suggested that the solute transport capability of intestinal tissue might be localized in the brush border boundary of the mucosal epithelial cell, those cells acting to concentrate solute which could then diffuse to the lamina propria region of the villi and hence have access to the circulatory system. This concept was strengthened with the development and use of unidirectional influx chambers which allowed the characteristics of solute entry into the brush border to be probed in detail over short time intervals before significant complications arose due to backflux or fluxes across other cell boundaries (Schultz *et al.*, 1967).

Together, the various experimental approaches described above focused attention on the function of the columnar epithelial cell of the small intestinal mucosa as the basis for intestinal absorptive phenomena. Quite logically, the thrust of scientific effort in this area began to be directed at learning more about the molecular mechanism by which these cells accomplish active transport. Data obtained from techniques involving transmural and unidirectional flux measurements have been particularly important in helping establish conceptual models for energy coupling to the Na^+-dependent transport systems both for amino acids (Curran *et al.*, 1967) and for sugars (Goldner *et al.*, 1969) (see Figs. 1 and 2). The models are fundamentally similar to the coupling mechanism first suggested by Crane *et al.* (1960, 1965) in which energy inherent in the cellular Na^+ gradient was envisoned as the driving force for solute accumulation. The central premise of this ion gradient hypothesis is that solute accumulation is dependent on and a consequence of the ubiquitous transmembrane cellular Na^+ gradient. The magnitude and polarity of the Na^+ gradient are regarded as the primary determinants of the magnitude and polarity of the solute gradients established. While a great many of kinetic and inhibitor data are consistent with such a concept (see reviews by Schultz and Curran, 1970, and Kimmich, 1973), a recognized important test involves examination of transport capability in tissue where the epithelial cell Na^+ gradient has been artificially manipulated or even reversed in

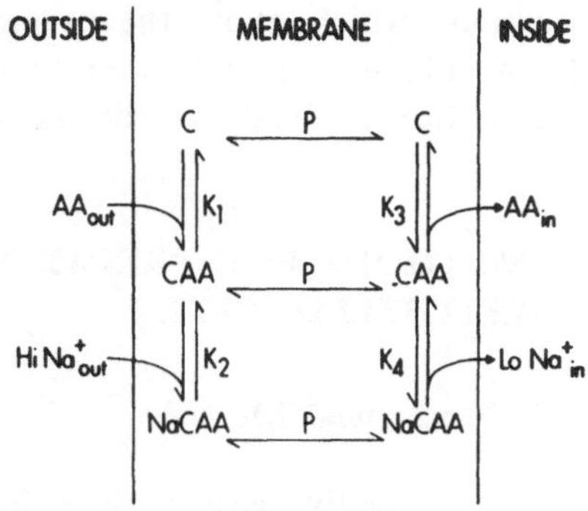

Fig. 1. Conceptual model for Na^+-dependent amino acid accumulation based on kinetic evidence derived from rabbit intestinal tissue by Curran *et al.* (1967). *C* represents an amino acid carrier, AA the amino acid, *P* a permeability coefficient for a specific carrier form, and *K* a dissociation constant for a given carrier species. Membrane translocation is regarded as rate limiting so that the carrier association–dissociation reactions are at equilibrium. Na^+ stabilizes the carrier–amino acid complex $(K_2 < K_1)$. High $[Na^+]_{out}$ favors ternary complex formation and low $[Na^+]_{in}$ favors dissociation by mass law considerations. Zero *net* flux of AA requires higher $[AA]_{in}$ than $[AA]_{out}$ as long as $[Na^+]_{out} > [Na^+]_{in}$.

direction in an effort to learn whether the transport systems exhibit the predicted changes in rate and direction of net solute flux. The latter test is exceedingly difficult to apply to preparations of intact tissue because of the tissue complexity and the presence not only of the epithelial cells but also of connective tissue, submucosa, several muscle layers, and relatively large extracellular spaces in the lamina propria region. It is impossible to manipulate tissue Na^+ concentrations and be certain of the precise change accomplished in epithelial cell Na^+. For this reason, development of

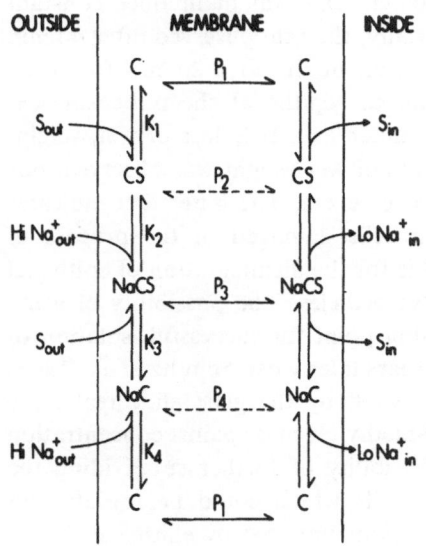

Fig. 2. Conceptual model for Na^+-dependent sugar accumulation based on kinetic evidence derived from rabbit intestinal tissue by Goldner *et al.* (1969). In this situation, the primary effect of Na^+ is to generate a ternary complex with greater mobility than either binary complex $(P_3 \gg P_2$ or $P_4)$. Low $[Na^+]_{in}$ favors dissociation to a binary complex with poor mobility. Further dissociation to free carrier is necessary before the carrier can return to the outer surface. Again, zero *net* flux of solute (sugar) requires $[S]_{in} > [S]_{out}$ as long as $[Na^+]_{out} > [Na^+]_{in}$.

methods which allow the preparation of intestinal epithelial cells free from submucosal and other tissue elements is important to the further elucidation of details of the transport mechanism.

2. METHODS FOR PREPARING ISOLATED INTESTINAL EPITHELIAL CELLS

2.1. Mechanical Methods

The relative ease with which intestinal mucosal epithelium can be separated from underlying submucosal elements by various scraping techniques quite logically led some investigators to use such techniques as a starting point for preparing suspensions of free epithelial cells. One of the earliest attempts of this type was reported by Dickens and Weil-Malherbe in 1941. They scraped the mucosal layer away from the submucosa with the edge of a glass microscope slide and studied the biochemical activity of the mucosal scrapings which contained epithelial cell sheets as well as individual cells, small cell clumps, fragmented villi, and cell organelles including free nuclei. They were the first to note the very high aerobic glycolytic capability of small intestine mucosa and demonstrated that the mucosal glycolytic rate nearly matched the rate for the intact small intestine. They also noted, however, that it was very difficult to incubate the preparation under conditions in which the glycolytic rate was maintained constant for any significant period of time. Typically, the rate observed after 40 min of incubation was less than half that seen in the first 20 min following preparation. During that same interval, the epithelial sheets became extensively fragmented and underwent as much as a 50% loss of wet weight. When intact tissue was incubated, no loss of wet weight was observed, nor was the loss of glycolytic activity quite so severe. The latter fact indicates that proteolytic activity released from cells damaged in the process of mechanical scraping might be responsible for the disintegration of epithelial sheets and loss of glycolytic activity. Nevertheless, the possibility of using scraped sheets of epithelium as a starting point for successful isolation of free cells was given some credibility 25 years later when Schultz *et al.* (1966) demonstrated that epithelial sheets prepared by the glass slide technique could accumulate both alanine and 3-*O*-methylglucose against concentration gradients. Their work indicated the feasibility of further subdividing the tissue into a population of individual cells which could be rapidly and reproducibly sampled without the restrictions imposed by epithelial sheets.

The latter preparations provide only single experimental samples in which determination of intracellular solute is complicated by the presence of rather large extracellular water spaces.

Despite this tantalizing possibility, the use of scraped mucosal preparations for epithelial cell isolation has not been a productive avenue of approach. Numerous investigators have reported the difficulty of fractionating mucosal preparations due to aggregate formation induced by the large quantities of mucus released at the time of scraping (Porteus and Clark, 1965; Clark and Porteus, 1965; Sjostrand, 1968). Aggregation of nuclei and microvillar sheets and formation of gels which resist centrifugal fractionation are particularly troublesome. Clark and Porteus (1965) reported success in isolating epithelial cell ghosts by gentle homogenization of mucosal scrapings in Krebs-Ringer phosphate containing 6% dextran followed by filtration and centrifugation. The cells retained all of the homogenate DNA and certain membrane-bound enzymes such as sucrase and succinic dehydrogenase, but had lost large portions of more soluble cellular material including 50% of the protein and RNA, and over 90% of certain glycolytic enzymes including lactic dehydrogenase and phosphoglucoisomerase. The ghosts did not respire unless cofactors were added and under no circumstances was glucose consumption or lactate production observed.

The most concerted efforts to isolate epithelial cells by purely mechanical methods of preparation have been reported by Sjostrand (1968) and Harrison and Webster (1969). Sjostrand (1968) developed several methods which depend on everting the intestine on a glass rod which is then placed in the chuck of a laboratory stirrer motor and rotated in a bathing medium held in a graduated cylinder. Rapid acceleration and deceleration in a large volume of 0.9% NaCl were found to be an effective way to remove mucus. If pressure was applied to the mucosal surface of the slowly rotating intestine, epithelial material could be induced to peel off in large sheets. The degree of contamination of the epithelial sheets with connective tissue and villi cores depended on the pressure applied and speed of rotation. A mechanical device was described (see Fig. 3) which can be used for applying pressure of any desired force to a plexiglas cylinder held against the side of the intestinal wall as the segment is moved longitudinally along the cylinder's lip. The largest sheets were obtained at 5000 rpm with no pressure applied, but pressure was required for the best yield of material and for freeing cells in the crypt region of the intestinal wall. Fractions corresponding to cellular clumps of various size, could be prepared by differential centrifugation, and the cells could be dispersed by discharging

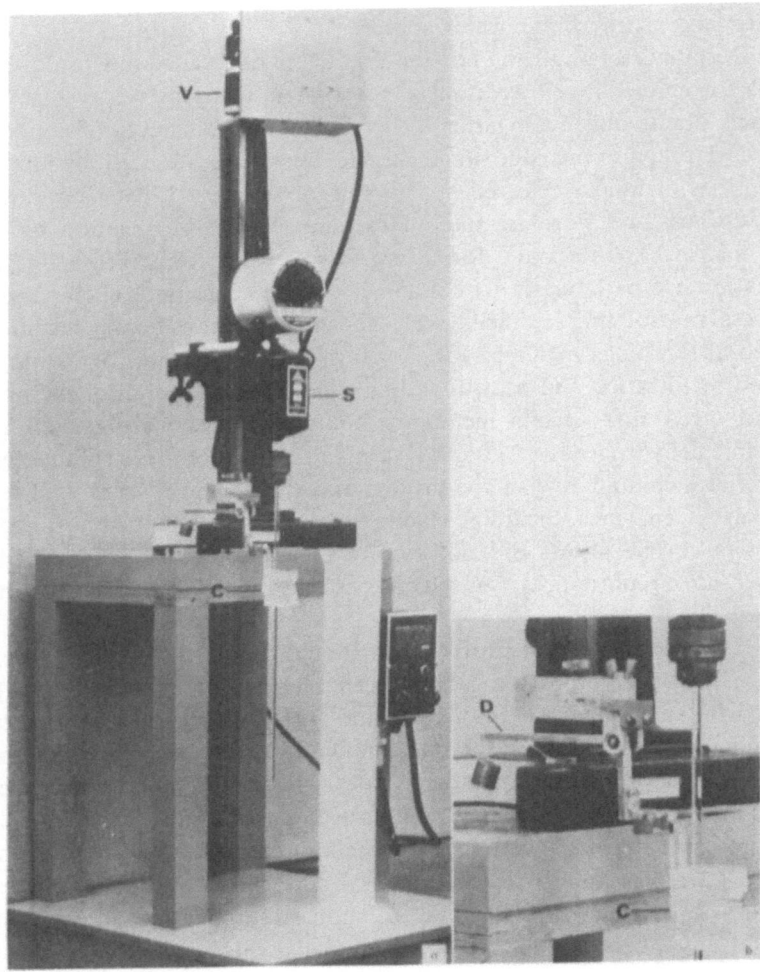

Fig. 3. Device for mechanically stripping epithelial layer from small intestinal tissue according to the procedure described by Sjostrand (1968). The intestine is everted on the lucite rod and rotated with the aid of a small stirring motor (S) while mechanical pressure is applied to the mucosal surface by the lucite block (C). A servomotor (V) moves the entire stirring assembly vertically in order to bring different regions of the intestine in contact with the lucite block. Tension applied to the gut wall can be varied by adjusting the position of the weight (D) on the lever arm. Photograph reproduced from Sjostrand (1968) with permission.

clumps washed free of mucus through a 24-gauge hypodermic needle. If hypertonic medium containing 0.2 M sucrose, 0.9% NaCl, and 2% dextran was employed, the cells were found to retain their columnar shape, while eliminating sucrose tended to yield rounded cells.

Harrison and Webster (1969) have used an approach which depends on mechanical vibration of an intestinal segment in order to free the epithelial cells. A segment of intestinal tissue is everted over a steel rod fastened in a Vibra-Mix motor and the villus cells are released by vibration at 100 cps for 30 min at an amplitude of 2 mm. A 5-min vibration is employed before cell collection is initiated to aid in removing mucus and easily detached cells which might represent a more senescent population removed from near the villus tips. Following removal of the cells from the lateral walls of the villus, the crypt cells can be released by distending the gut segment with air pressure, attaching a weight to the distal end, and continuing vibration for 10–20 s. Subsequent work by the same authors (Webster and Harrison, 1969) indicated that cells can be removed from relatively specific areas of the villus by proper choice of the vibration interval. Cells near the villus tips are removed most readily, while longer vibration periods remove progressively those cells situated more toward the crypt region. Crypt cells are not removed unless the gut wall is distended.

Iemhoff *et al.* (1970) have attempted to characterize the metabolic capability of cells isolated by the Sjostrand (1968) and Harrison and Webster (1969) techniques (see Table III). The Sjostrand method was found to yield cells which tend to be swollen, contain numerous vacuoli, and exhibit rather low rates of lactate production (120 nmol/mg protein · h). Mitochondria prepared from such cells exhibit poor coupling characteristics, as indicated by low P/O ratios. Cells prepared by the Harrison and Webster techniques are less swollen, exhibit much higher glycolytic capability (600 $n$mol/mg protein · h), and yield mitochondria with better respiratory control. The vibration procedure was judged superior to rotation with mechanical pressure for cell preparation, although no attempts were made to examine solute transport capability of the cells prepared by either method.

2.2. Chemical and Enzymatic Methods

A number of cell isolation procedures which depend on treatment of intact tissue with various chelating agents and/or enzymes were reported in the mid-1960s. The earliest report by Stern and Reilly (1965) described incubation of intestinal segments filled with a sucrose–saline medium containing trypsin-pancreatin. Following 15 min at 35°C, epithelial cells

Table III. Characteristics of Intestinal Epithelial Cells Isolated by Various Techniques

Procedure	O_2 consumption (μl O_2/mg protein $\cdot$ h)	Lactate production (μmol/h $\cdot$ mg protein)	CO_2 production (μmol/h $\cdot$ mg protein)	Transport capability		Comments	Ref.[a]
				Sugars	Amino acids		
Mucosal scraping + centrifugation	None	None	N.D.[b]	N.D.	N.D.	Cell ghosts	1
Trypsin-pancreatin + mechanical pressure	5–10	0.3–0.35	N.D.	Variable	N.D.	Sugar transport not Na^+ dependent	2
Lysozyme treatment of minced intestine	3.3	N.D.	2.1	4-fold gradients for 3-OMG	4-fold gradients for Tyr	Transport only 40% sensitive to 3.3 mM DNP	3
Hyaluronidase treatment of everted sacs	21.6	N.D.	N.D.	+	N.D.	Sugar transport capability inferred with by sugar-induced increase in O_2 consumption	4
EDTA treatment with mechanical agitation	12.5	N.D.	N.D.	No active accumulation	<2-fold glycine gradients	Glycine uptake not sensitive to DNP, anoxia, or lack of Na^+	5
Mechanical pressure applied to surface of rotating intestine	N.D.	0.12	N.D.	N.D.	N.D.	—	6
Vibration of everted sac in media containing EDTA	N.D.	0.6	0.19*	N.D.	N.D.	Very low energy charge[c] = 0.45: AMP + ADP = 6 ATP	7
Citrate + hyaluronidase treatment + mechanical pressure	N.D.	N.D.	N.D.	N.D.	<2.5-fold gradients	Gradients estimated assuming cell H_2O = 5 × dry wt	8
Hyaluronidase + mechanical agitation	1.35	0.3–0.5	0.5	4- to 8-fold gradients	4- to 8-fold gradients	Na^+ dependence well characterized Energy charge = 0.75	9

[a] References: (1) Clark and Porteus (1965), (2) Stern (1966), (3) Huang (1965), (4) Perris (1965), (5) Sognen (1967), (6) Iemhoff et al. (1970), (7) Iemhoff et al. (1970), (8) Reiser and Christiansen (1971), (9) Kimmich (1970a,b).

[b] Not determined.

[c] Energy charge = $(2\text{ATP} + \text{ADP})/[2(\text{ATP} + \text{ADP} + \text{AMP})]$, where AMP, ADP, and ATP are the concentrations of the nucleotides.

were released from the mucosal wall by gently pressing the serosal surface
with the fingers. The cell suspension was collected in a syringe, more
trypsin-pancreatin was added, the suspension was discharged through a
14-gauge needle, and cells were collected by centrifugation for 5 min at
500g. After two washings in fresh medium to remove the enzymes, the cells
were ready for use. In a preliminary report, Harrer *et al.* (1964) had reported that exposure of the intestine to trypsin for periods longer than
15 min led to poorer cell viability than when exposure time was of a shorter
duration. In an effort to avoid the use of proteolytic treatment, Stern and
Jensen (1966) used 27 mM sodium citrate in calcium/magnesium-free,
phosphate-buffered saline in place of trypsin treatment. Cells were again
released by manually pressing the serosal wall. DNase was added to the
suspension to help reduce viscosity and the cells were characterized by a
number of biochemical criteria (Stern, 1966). The preparation exhibited
both respiratory and glycolytic activity which was sensitive to the addition
of 5 mM iodoacetate. The best respiratory activity occurred in the presence
of succinate (8 μl O_2/mg protein $\cdot$ h) and was not affected by the addition
of ADP, Ca^{2+}, Mg^{2+}, NAD, glucose, or glutamate, indicating that the cell
membranes remained intact. The cells glycolyzed at a rapid rate (300–350
nmol lactate/mg protein $\cdot$ h) and exhibited only a modest Pasteur effect,
consistent with the observations of Dickens and Weil-Malherbe (1941) with
mucosal scrapings. An attempt was made at estimating glucose uptake
by the preparation; glucose gradients ranging from 2- to 60- fold were
reported. However, in no case was formation of the gradient inhibited
by removal of Na^+ from the medium, which casts doubt on the validity
of their technique. It is known that active intestinal sugar transport is
absolutely dependent on the presence of Na^+ in the bathing medium.

Another approach using enzymes was described by Huang (1965).
This involved incubating an intact intestinal segment for 10–15 min at 36°C
after filling the venous system with a 0.25% lysozyme solution. After the
initial incubation the lumen was filled with the lysozyme solution, and
incubation continued for 30 min. The entire segment was then minced and
incubation was continued another 10 min with vigorous mechanical stirring.
Finally, the mince was filtered through a double layer of gauze and centrifuged at 500g for 2 min. Huang did not characterize the pellet obtained,
but it almost certainly contained whole and fragmented villi and fragments
of connective and muscle tissue considering the fact that the entire intestine
was initially fragmented rather than just the epithelial layer. Huang was,
however, one of the first to examine the solute transport capability of an
intestinal cell preparation. He reported that small gradients of tyrosine

and 3-*O*-methylglucose could be established which tended to dissipate after 10 min of incubation.

In 1967, Sognen reported a method which depended on the release of epithelial cells induced by 2 mM EDTA in a calcium/magnesium-free Krebs-Ringer phosphate buffer. An everted sac was incubated with the chelating agent for 1 h at 4°C, then subjected to vigorous shaking in order to detach epithelial sheets. The transport capability of this preparation was also examined. No active accumulation of 3-*O*-methylglucose could be observed and only a small (less than 2-fold) gradient of glycine was detected; this could not be inhibited by the uncoupler dinitrophenol. The reported characteristics of cells prepared by each of the methods mentioned above are summarized in Table III.

When we first began considering techniques for preparing isolated intestinal epithelial cells in my laboratory, our first inclination was to try to separate intact cells from a suspension of mucosal epithelium scraped free of the intestinal wall in order to avoid the rather vigorous mechanical techniques otherwise required to release such cells. It quickly became apparent that aggregate formation due to the large amounts of mucus in such preparations would have to be circumvented if adequate separation of cells from debris, nuclei, and tissue fragments were to be achieved. Indeed, the quantity of mucus released is frequently so significant that the scraped mucosal preparation forms a gel which resists centrifugal fractionation and cannot even be readily transferred from one vessel to another. We found that hyaluronidase (1 mg/ml) added to the scraped suspension prevents gel formation and allows ready centrifugation of individual cells and cell clumps. Due to extensive clumping, we decided to separate cells further by exposing the hyaluronidase-treated preparation to various enzymes or chelators, some of which had been employed by other investigators as described earlier. Each was tried alone or in combination with one or more of the others and every method was found to yield free cells which could be readily separated by centrifugation. However, the metabolic capabilities of such preparations were found to deteriorate very readily, as time elapsed after preparation. Microscopic studies indicated that, while large numbers of intact cells were initially separated, the cell population autolysed extremely readily, and intact cells were observed to lyse spontaneously during a short period of examination under the microscope. We predicted that cells damaged during the scraping procedure might be releasing large amounts of proteolytic enzyme which then could attack and digest the intact cell membranes leading to lysis. For this reason, bovine serum albumin was included in the saline medium during scraping, treatment with hyaluronidase, and cell separation in order to provide a substrate for the suspected enzyme and prevent cell deterioration. The added albumin was moderately successful in delaying autolysis, but no conditions were found which eliminated extensive autolytic activity and lysis of cells.

At this point, we turned to the use of transverse slices of intestinal tissue as a possible alternative to the use of isolated cells for transport studies. Indeed, such tissue "rings" actively accumulate amino acids and sugars, as reported by Agar *et al.* (1954, 1956) and Crane and Mandelstam (1960). More significantly, we noted that the slices could be incubated for extended intervals (4 h or more), with no perceptible disintegration or autolysis of the epithelial cell layer. This observation suggested that cells resistant to self-degradation might be prepared if they could be loosened from the gut wall by techniques other than scraping that would minimize release of proteolytic enzymes. In light of our success with hyaluronidase for removing mucus, we decided to treat intact intestinal tissue obtained from chickens with hyaluronidase in hope of loosening the epithelial cells from their extracellular "fuzzy coat" on the mucosal intestinal surface. We found that incubation with hyaluronidase (1 mg/ml) in a modified Krebs-Ringer phosphate buffer containing serum albumin (1 mg/ml) releases a few cells spontaneously, and we found that large numbers can then be freed by gently stirring the intestinal segments with the tip of a plastic pipette. The suspension of material released is then poured through nylon stocking material to aid in removing large aggregates of particulate material, undigested mucus, fragmented villi, and large cell clumps. Cells in the filtrate can be readily sedimented at 100g for 1–2 min. Cells in the pellet tend to aggregate readily due to residual mucus, but they can be dispersed by adding fresh medium free of hyaluronidase, drawing the medium up in a plastic pipette, and blowing it out while directing the stream of liquid toward the cell pellet. Repeated refluxing in this manner leads to a suspension which can be easily drawn up in a small-bore pipette. Centrifugation and rewashing in fresh medium can be used to remove hyaluronidase carried down in the original pellet. However, we have shown that cell suspensions incubated in the presence of hyaluronidase have essentially the same metabolic and transport capability as in the absence of the enzyme. Apparently no deleterious effects on the cell membrane are induced by the hyaluronidase, which indicates the value of using it in preference to proteolytic enzymes or chelators which can induce undesirable changes in membrane function.

An intestine taken from 6- to 8-wk-old chickens yields approximately

1.5–2.0 ml of cell pellet when packing is performed at low centrifugal forces (100–200g). Total protein content of the cell pellet is typically 60–80 mg per milliliter packed cells. It is necessary to carry out all procedures in plastic labware and to use plastic pipettes in order to avoid extensive fragmentation of cells. Other investigators (Sjostrand, 1968) have reported similar experience. Each step after the initial treatment with hyaluronidase is performed at 4°C.

Apparently the procedure described above will not be equally successful with gut obtained from every animal species. In repeated attempts at using intestinal tissue from rats, we have obtained very poor yields of cells which possess minimal ability for active accumulation of sugars and amino acids. Reiser and Christiansen (1971) have reported a preparation of rat intestinal cells using the citrate-chelation–mechanical-pressure method of Stern and Jensen (1966) modified to include hyaluronidase during the initial incubation of intact intestine. Cells prepared by their technique retain a weak capability for active accumulation of amino acids, but their capability is poor in comparison to that of cells isolated from the chick intestine, as will be described in Section 4. Perris (1965) has described a procedure utilizing hyaluronidase for preparing epithelial cells from rat intestine which is very similar to the one we described for chick intestinal cells, but unfortunately the metabolic and transport capability of his preparation was not adequately defined. Oxygen consumption was examined as an index of metabolic activity, but no measurements of solute transport capability were reported by Perris. A comparison of the characteristics of intestinal epithelial cells prepared by each of the methods mentioned above is given in Table III. Most of the data to be described in the remaining part of this chapter have been obtained using cells prepared from chick intestine by the hyaluronidase procedure, although certain comparisons with cells prepared by other techniques are included in order to allow an evaluation of the relative merit of various techniques.

3. CHARACTERIZATION OF ISOLATED INTESTINAL EPITHELIAL CELLS

3.1. Microscopy and Dye Exclusion Methods

Microscopic examination of an epithelial cell suspension prepared by the hyaluronidase procedure shows that a large fraction of the total population exists in small clumps and groups consisting of two to ten or more

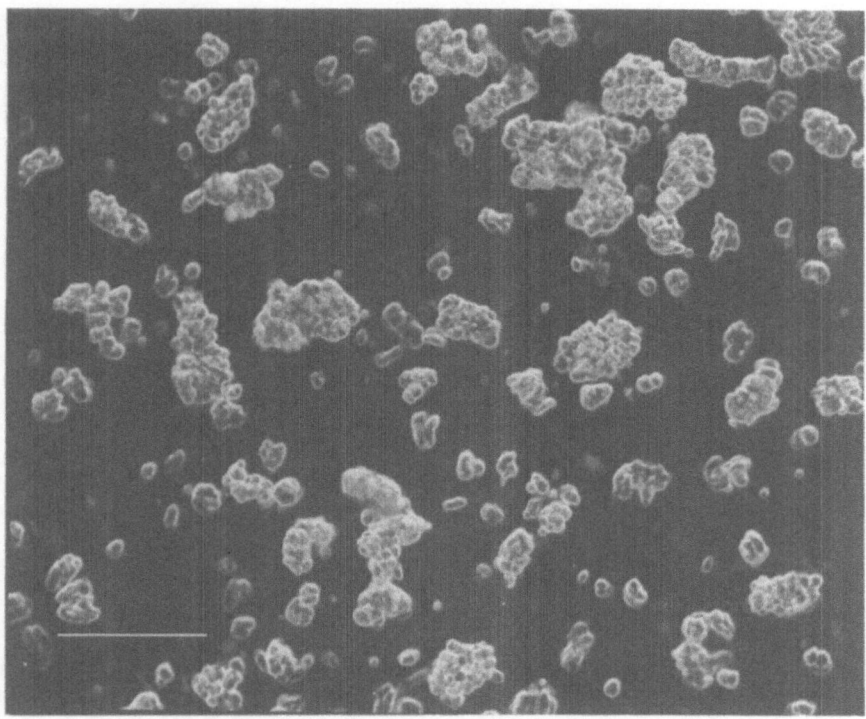

Fig. 4. Low-power phase contrast microscopy of a cell suspension prepared from chick intestine by the hyaluronidase procedure described in the text. Bar (lower left) = 100 μm.

individual cells. Extensive refluxing of the suspension through a small-diameter pipette orifice tends to dissociate the larger clumps to produce more uniformly sized groups of cells, but it is impossible to disperse the population completely to individual cells (see Fig. 4). Extensive refluxing is undesirable in that it produces a population with poor metabolic and transport capability, which apparently indicates cell damage due to the mechanical forces required for complete dissociation. Furthermore, a cell population which has been refluxed simply the minimal amount required for dispersion of the centrifuged pellet can be readily drawn up in micropipettes and sampled quite reproducibly ($\pm 5\%$), as a determination of cell protein in replicate samples taken from a single suspension will show (Table IV). In light of the ease of sampling and the excellent functional capability of such preparations (see Section 3.2), we routinely make no effort to disperse a cell pellet beyond that point necessary for easy handling with a pipette of small tip diameter.

Table IV. Protein Determinations of Replicate Samples Taken by Micropipette from a Suspension of Isolated Intestinal Cells

	mg protein/200 μl cell suspension
	19.9
	21.0
	20.0
	19.3
	20.5
Mean	20.1 ± 0.2

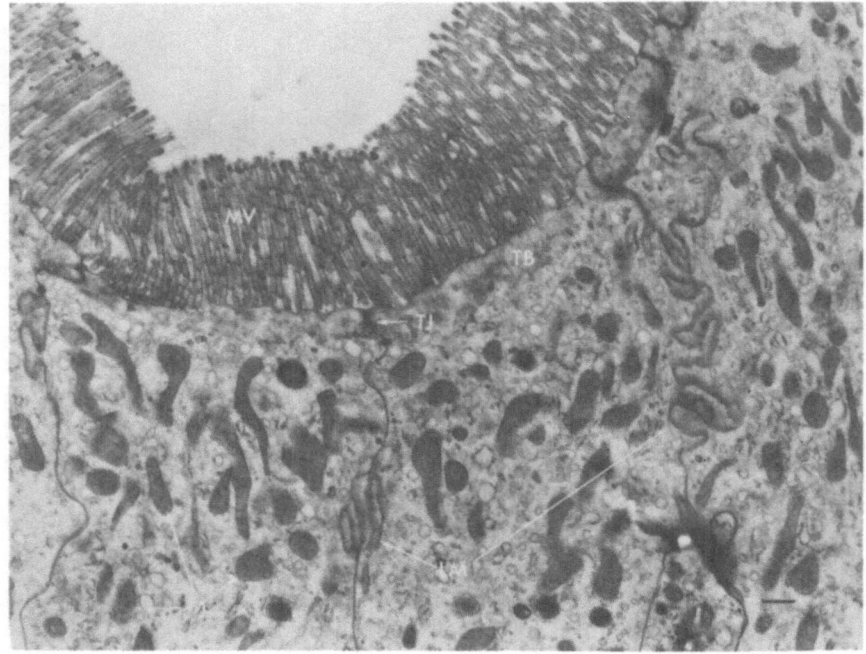

Fig. 5. Electron micrograph of the mucosal epithelium of small intestinal tissue from the chicken. MV, Microvillus region; M, mitochondria; TJ, tight junction; TB, terminal bar region. Note the degree of interdigitation of lateral membranes (LM) of adjacent cells. Bar (lower right) = 1 μm.

A further indication that caution should be exercised in attempts to fully disperse the cell population is apparent with even a brief consideration of intact intestinal epithelium and the nature of *in situ* cell attachment. Figures 5 to 7 are electron micrographs of the mucosal epithelium of chick jejunum taken at different degrees of magnification. At the lowest magnification (Fig. 5), the rather tortuous interdigitation of adjacent lateral cell boundaries is unmistakable. While these lateral boundaries do not appear to have intercell attachment sites over much of their extent, scattered desmosomal attachments are nevertheless apparent. Furthermore, note that without exception there are tight junctional complexes between each pair of cells on the lateral boundary just below the luminal brush border (microvillus) region. At higher degrees of magnification (Figs. 6 and 7), the fusion of adjacent cell membranes in the tight junction region can be ascertained. These areas represent a band of attachment which completely encircles the columnar epithelial cell. The desmosomes which appear somewhat below

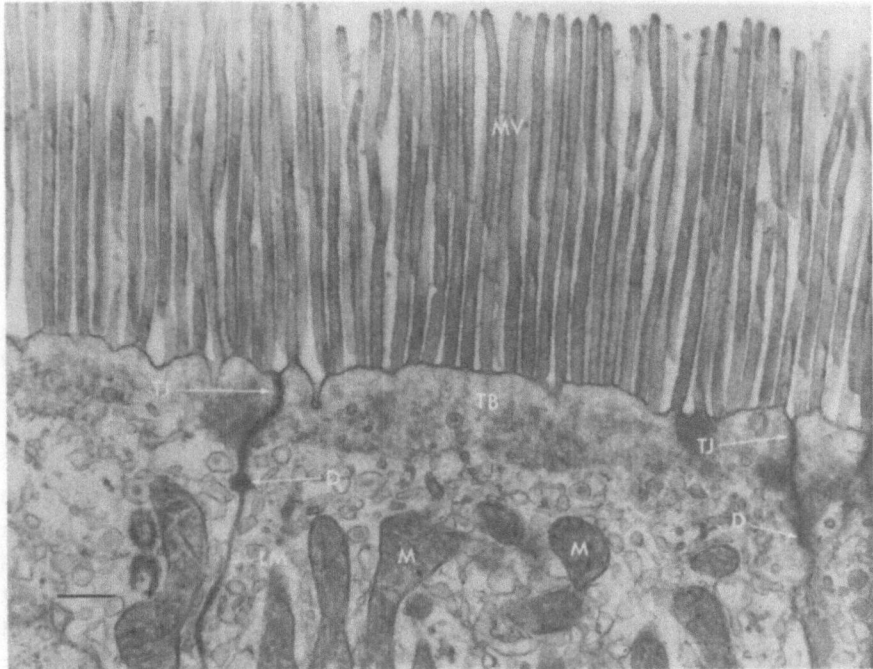

Fig. 6. Electron micrograph of the lateral membrane junctional complexes between adjacent mucosal epithelial cells in chicken small intestine. D, Desmosome. Other lettering as designated for Fig. 5. Bar (lower left) = 1 μm.

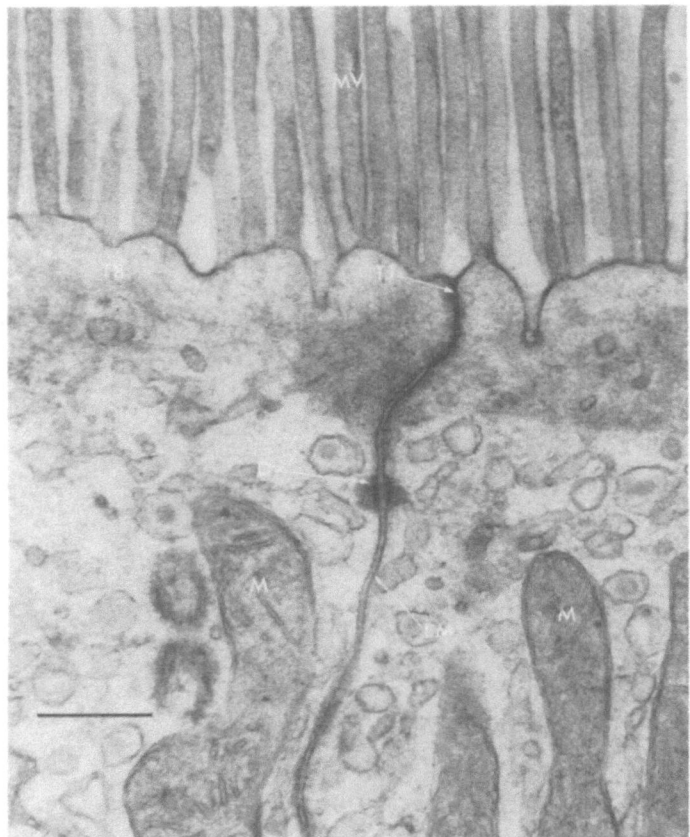

Fig. 7. High-power electron micrograph of membrane junctional complexes between adjacent intestinal mucosal epithelial cells. Lettering designations are the same as indicated for Figs. 5 and 6. Bar (lower left) = 1 μm.

the tight junctional region, on the other hand, represent only specific points of cell attachment. Obviously, such "spot welds" could serve to make the lateral boundary interdigitation somewhat more effective in holding adjacent cells together. In order to completely free cells from their neighboring cells it is thus necessary to disrupt not only lateral membrane involutions but also the firmer sites of attachement represented by desmosomes and tight junctional complexes. It is not surprising therefore that a high percentage of cells in an epithelial preparation exist in small clumps or groups of cells; nor should it be unexpected that full separation of individual cells might lead to membrane damage and poor functional capability. In fact, detailed examination of electron micrographs of cells completely freed of their neighbors frequently indicates accessory material attached at the original tight junction area which apparently represents membranous material torn from an adjacent cell. It seems likely that completely freed cells can be obtained only at the expense of neighbor cells as the tight junctional complex is disrupted. The important goal then of the preparation procedure is not to produce a suspension of individual cells as much as it is to free groups of cells which retain their physiological metabolic and functional capability and which can be readily adapted to various laboratory procedures for examining that capability.

Because of the nature of the cell population produced by the hyaluronidase procedure, exact quantitation of cell numbers in a given suspension is not particularly easy. Counting cells in a hemocytometer is complicated by the presence of small groups of cells of uncertain number which necessarily allows only approximate counts. Our best estimates by this procedure indicate that 1 mg of cell protein represents approximately 1 million cells. Cell protein can be conveniently determined using the biuret assay (Gornall et al., 1949) if 1% Triton is included in the reagent to allow full solubilization of the cellular material. Any residual turbidity after color development indicates that too much cell suspension was employed and such samples must be assayed in higher dilution. Blanks and standards must include the identical amount of saline medium as will be present in the cell samples, since we have found that salt concentration and ionic strength influence the color-forming reaction.

Despite the difficulty in exactly quantitating cell number, we have found that functional capability expressed in terms of amount of cellular protein yields data with a high degree of reproducibility from one preparation to the next. Apparently the cells prepared from various animals are uniform enough in size so that the protein values reliably reflect cell numbers. In situations where animals of different age will be employed for comparative purposes, DNA content of the suspension might be an even better standard, although we have not yet evaluated this possibility.

As a result of the difficulty in quantitating cell numbers accurately, it is also difficult to obtain a reliable estimate of cell viability by dye exclusion methods. We have estimated that approximately 80% of cells prepared by hyaluronidase treatment remain viable using trypan blue exclusion as an index of viability. Given the problems in counting cell groups, this figure represents only an estimate, although it is in agreement with values obtained by other investigators for other preparative techniques (Stern and Reilly, 1965; Reiser and Christiansen, 1971). However, comparison of the metabolic and functional capability of these preparations with that exhibited by the hyaluronidase preparation indicates that the latter procedure yields a much superior cell population (see next section). We therefore are uncertain of the usefulness of dye exclusion methods as indicators of cell integrity. This uncertainty is reinforced by recent observations (Barrett, 1974; Barrett and Coleman, 1973) which indicate that even after osmium tetroxide fixation, cells of various kinds exclude trypan blue for several hours. This was observed despite the fact that the osmium fixation produces nonviable cells and is believed to render cell membranes permeable to large molecules. In light of these observations, we have placed more emphasis on metabolic and transport capability as indices of cell integrity and function. Nevertheless, it is important to keep in mind that a certain fraction of any population of intestinal epithelial cells in all likelihood represents damaged cells with limited or no functional capability. This likelihood requires consideration during the interpretation of data obtained with such preparations (see review by Kimmich, 1973, Sections VIII and IX).

3.2. Metabolic Characterization

Several of the epithelial cell preparations described by other investigators have been studied in terms of their respiratory and glycolytic capability (Stern and Reilly, 1965; Perris, 1965; Stern and Jensen, 1966; Iemhoff et al., 1970) as a means of learning the extent to which normal metabolic activity is retained by the isolated cells. Although respiratory or glycolytic capability was observed in each instance, in those cases where the activity was monitored as a function of time it was noted that linear rates of oxygen consumption or lactate production could be demonstrated only for relatively short periods of time (30 min or less). This phenomenon was reminiscent of the data obtained by Dickens and Weil-Malherbe (1941) with scraped mucosal epithelial preparations, and raised the possibility that extensive autolytic disintegration may have been a detrimental influence in each of the cases studied. Our own data obtained with scraped epithelial material, alluded to earlier, had indicated the same problem.

In contrast to these results, the metabolic activities of intestinal epithelial cells prepared from chicken by the hyaluronidase method exhibit linearity for periods of up to 2 h. Figure 8 shows the production of lactate and carbon dioxide from glucose as a function of time when an aliquot of cells is incubated in a saline medium similar to that employed for isolation of the cells, but supplemented with glucose and 10 mg/ml of bovine serum

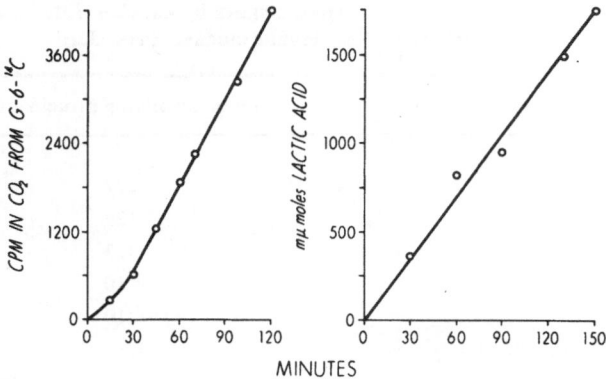

Fig. 8. Carbon dioxide and lactic acid production from glucose by isolated intestinal cells prepared from chicken tissue by the hyaluronidase procedure. Each vessel contained 10 mg protein for determination of $^{14}CO_2$ evolved from $[6\text{-}^{14}C]$glucose, and 2.3 mg protein in those cases where lactic acid production was determined. Reprinted from Kimmich (1970a).

albumin. We have noted a tendency for the periods of linear metabolic activity to be somewhat more prolonged in the presence of elevated concentrations of serum albumin. This is consistent with Dickens and Weil-Malherbe's (1941) observation that epithelial sheets incubated in serum do not undergo the extensive disintegration and weight loss noted in the absence of serum. In a related observation, Perrin (1965) demonstrated that the respiratory activity of rat intestinal cells was doubled when bovine serum albumin was included in the incubation medium.

Figure 8 indicates that glycolytic activity by the chick intestinal cells produces about 300 nmol of lactate per milligram of cell protein per hour. Comparable values have been established for cells isolated by the citrate chelation method (Stern and Jensen, 1966) and the mechanical vibration method (Temhoff et al., 1970). In contrast, Iemhoff et al. (1970) found that cells prepared from epithelial sheets stripped by mechanical pressure during rotation of an everted gut sac glycolyzed relatively poorly (120 nmol/h · mg protein) (see Table III). Higher values are usually accepted as indicative of an undamaged cell population, in light of the well-documented fact that intestinal tissue ordinarily exhibits high rates of aerobic glycolysis (Dickens and Weil-Malherbe, 1941). In fact, values for lactate production near 500 nmol/h · mg protein compare very favorably with glycolytic rates determined for ascites cells, which are usually considered to have an exceptionally rapid aerobic glycolytic capability (Wu and Racker, 1959).

Table V. Lactate Production from Various Sugars by Chicken Intestinal Epithelial Cells Prepared by the Hyaluronidase Procedure

Substrate	nmol lactate/mg protein · h
Glucose	277
Mannose	125
Fructose	123
Galactose	<10[a]
Sorbose	<10
3-O-Methylglucose	<10
Ribose	<10
Sucrose	85

[a] Under the conditions employed, 10 nmol lactate per milligram protein per hour represents the lower limit of sensitivity for the assay procedures under the conditions employed. Data from Kimmich (1970a).

While glucose supports the highest rates of lactate production, a number of other sugars have also proven to be good glycolytic substrates. As shown in Table V, mannose and fructose each support lactate production at about half the rate produced by glucose. Sucrose was found to support a still slower rate, although the experiment was performed in 20 mM tris buffer, which has been reported to inhibit sucrase activity. Nonmetabolized sugars such as 3-O-methylglucose, sorbose, galactose, and ribose supported little or no lactate production, as expected.

An attractive advantage of using isolated cells in contrast to intact tissue is the ease with which cell suspensions can be utilized for determining various types of metabolic capability. Fluid cell suspensions can be readily employed for manometric measurements of several types as well as in conjunction with polarographic techniques utilizing a platinum electrode for direct determination of respiratory activity. We have made use of several techniques ordinarily employed with preparations of isolated mitochondria for characterizing the metabolic capability of epithelial cells prepared by the hyaluronidase method. If radioactive glucose is used as substrate and carbon dioxide is collected during incubation, it can be shown that the chick cells produce carbon dioxide at linear rates for periods of time up to 2 h (Fig. 8). Typically 50–60 nmol of CO_2 is produced per hour per milligram protein, which is equivalent to the complete oxidation of about 20 nmol of pyruvate (or lactate). Apparently a very high percentage of the

3-C compounds produced from glucose appear as lactate rather than being oxidatively metabolized. Short-term experiments employing an oxygen electrode confirm this likelihood, indicating a respiratory activity equal to 60 nmol oxygen consumed per hour per milligram protein. This value corresponds to the total oxidation of about 25 nmol of pyruvate per hour per milligram protein, in good agreement with values obtained from $^{14}CO_2$ measurements.

Both respiratory activity and glycolytic activity of the isolated cells are responsive to the presence of various metabolic inhibitors, as shown in Figs. 9 and 10. Uncoupling agents such as dinitrophenol induce an increase in both activities, as would be expected for any situation where metabolic activity is freed from the restraints ordinarily imposed by coupling to chemical energy production. On the other hand, if cellular energy production is closely matched to total energy expenditure, any agent which specifically inhibits an energy-dependent event would tend to preserve the cellular energy status and decrease metabolic activity. The magnitude of decrease should be an indication of the fraction of total energy turnover ordinarily related to the inhibited event. In this regard, the effects of ouabain, a specific inhibitor of monovalent ion transport, are particularly interesting. Ouabain produces more than 70% inhibition of lactate production and 50% inhibition of carbon dioxide production by the isolated epithelial cells, indicating that a rather high proportion of the total energy expended by such cells is used in order to maintain gradients of monovalent cations. Similar degrees of inhibition are achieved by simply removing K+ from the

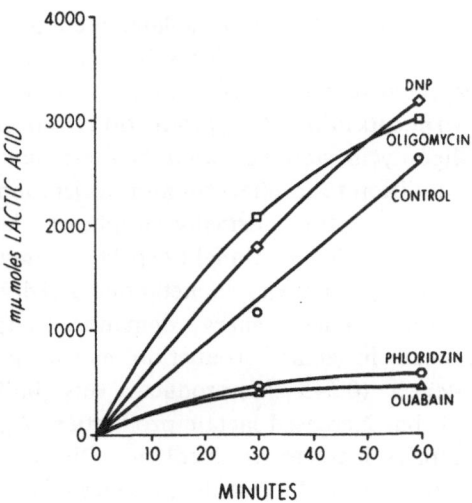

Fig. 9. Effect of various metabolic inhibitors on the production of lactic acid by isolated intestinal epithelial cells. Each vessel contained 4.5 mg of cell protein. From Kimmich (1970a).

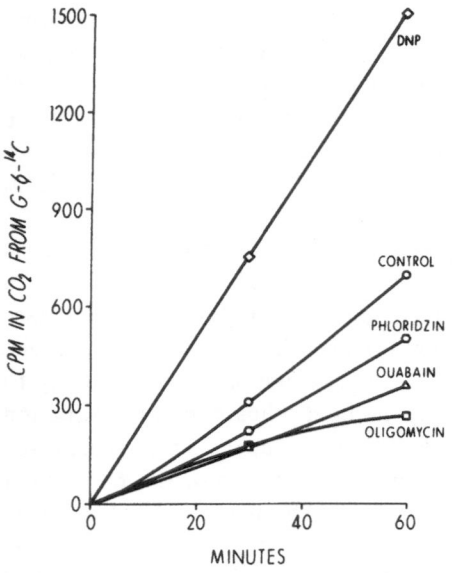

Fig. 10. Effect of various metabolic inhibitors on the production of $^{14}CO_2$ from [6-^{14}C]glucose. From Kimmich (1970a).

suspending medium, which would also inhibit the monovalent cation transport system. Racker has recently shown that ascites cells and certain other types of tumor cells also exhibit high rates of aerobic glycolysis which are unusually sensitive to inhibition by ouabain. Perhaps in these situations there is a degree of integration between glycolytic activity and the capability for rapid transport of monovalent cations. This interesting possibility will be discussed further in a later section.

Oligomycin, another inhibitor of monovalent ion transport, might be expected to exert effects similar to those induced by ouabain. Indeed, the two agents inhibit CO_2 production from glucose to nearly equivalent degrees. Oligomycin, however, inhibits a terminal step in oxidative phosphorylation in addition to its effect on monovalent ion transport. Therefore, if oxidative phosphorylation normally supplies a significant fraction of total cellular ATP expended, one might expect oligomycin to induce an adaptive increase in glycolytic activity as metabolic activity shifts to a nonoxidative pathway in order to meet energy demands. In fact, oligomycin causes a marked increase in lactate production, as shown in Fig. 9. Low concentrations of rotenone (0.5–5 μM) produce a very similar set of responses, i.e., decreased CO_2 but increased lactate production. Again, this apparently represents an adaptive increase in glycolytic activity in response to decreased oxidative production of ATP in the presence of the electron transport inhibitor.

Table VI. Comparison of Adenine Nucleotide Concentrations and Calculated Energy Charge Values for Intestinal Epithelial Cells Prepared by Hyaluronidase Treatment or Mechanical Vibration Techniques

Preparative procedure	Nucleotide concentration			Energy charge[a]
	ATP	ADP	AMP	
Vibration	67[b]	300	100	0.45
Hyaluronidase[c]	3.4[d]	0.52	1.11	0.73
+10 μM rotenone	0.14			
+5 μg/ml oligomycin	0.49			
+200 μM ouabain	3.57			
+200 μM phloridzin	3.45			

[a] For definition of energy charge, see footnote c of Table III.
[b] Relative values determined by Iemhoff et al. (1970) for cells prepared by the method of Harrison and Webster (1969). AMP value arbitrarily set at 100.
[c] Cells were prepared as described in the text and incubated for 10 min with or without the indicated inhibitor before ATP determinations were performed.
[d] All values for cells prepared by the hyaluronidase procedure are expressed as nmol nucleotide/mg cell protein.

Despite the rapid glycolytic activity induced by rotenone and oligomycin, the total cellular ATP level decreases, as shown in Table VI. In light of information regarding the control of glycolytic activity in other tissues, it seems likely that the decreased ATP levels or elevated ADP levels (or both) may be the important signals for stimulating glycolysis in these two situations. Consistent with this interpretation, ouabain, which markedly inhibits glycolytic activity, actually causes an elevation in ATP levels. This was expected since ouabain should prevent ATP expenditure due to monovalent ion transport, but not interfere with the various mechanisms of ATP synthesis.

We have also determined the ADP and AMP content of freshly prepared chick intestinal epithelial cells after 10 min of incubation in the saline medium ordinarily employed for metabolic and transport characterization. The values found are given in Table VI. The energy status of the cell population compares very favorably with that of intact tissues when expressed as the calculated energy charge (energy charge = $\frac{1}{2}(2ATP+ADP)/(ATP+ADP+AMP)$, where AMP, ADP, and ATP are concentrations of the nucleotides). The observed value of 0.73 is much higher than the value of 0.45 for cells prepared by the mechanical vibration method of Harrison and Webster (1970) as reported by Iemhoff et al. (1970). To our knowledge, that is the only other preparation of intestinal epithelial cells in which adenine nucleotide levels have been established. The values established by Iemhoff *et al.* are also given in Table VI. Note that ATP levels for cells prepared by vibration are only one-fifth those of ADP, and that even AMP levels exceed those for the nucleoside triphosphate. Chapman *et al.* (1971) have concluded that cells from a wide variety of tissues maintain an energy charge near 0.8 when they are in a fully functional state. Viability is maintained at energy charges as low as 0.5, but rapid loss of viability occurs when the energy charge falls below 0.5. Despite the metabolic activity demonstrated for the Harrison and Webster preparation (Iemhoff *et al.*, 1970), such peculiar adenine nucleotide ratios raise doubts as to the true functional capability of the isolated cells.

In considering the preceding statement, we were prompted to consider whether simple microscopy and metabolic characterization of a given cell population are very informative with respect to membrane integrity and functional status of the population. True, it is satisfying to demonstrate linear rates of metabolic activities of long duration, and expected effects of various types of inhibitors, but are these important attributes of intact viable cells? Recalling that crude cell-free homogenates prepared under the appropriate conditions will frequently exhibit linear sustained rates of lactate and carbon dioxide production which are responsive to various inhibitors, it seems fair to conclude that such capability, in itself, need not reflect the degree of success in preparing intact functional cells. In fact, Stern and Reilly (1965) reported that homogenization of an epithelial cell suspension prepared by their method did not alter the rate of lactate production by those cells. Clearly, additional criteria beyond metabolic activities are required to fully establish whether a given preparative procedure can produce cells useful for probing all aspects of the nature of the tissue from which they were derived.

4. EVALUATION OF TRANSPORT CAPABILITY

4.1. Methods: Filtration vs. Centrifugation Techniques

If an indication of membrane integrity can be accepted as a worthwhile goal for adequate characterization of a cell population, it seems logical to examine the active transport capability of the population as a reflection of the degree to which the membrane has retained its ability to maintain concentration gradients of solute. This seems particularly important when dealing with an epithelial preparation derived from the intestine, where the opportunity for cell rupture is increased as cell–cell tight junctions are disrupted. Aside from structural integrity, normal transport capability would be an indication of minimal alterations of those specific membrane components necessary to allow mediated entry as well as an expression of intact coupling devices for allowing metabolic energy to be partially conserved in the form of electrochemical solute gradients.

Despite the obvious value in ascertaining the transport capabilities of epithelial cell preparations, we are aware of only three cases in which attempts were made to assess this function. The best solute gradients were claimed for the Stern and Jensen (1966) preparation, but no Na^+ dependence could be demonstrated even though Na^+ is known to be essential for active transport in the intact tissue. Cells prepared by the EDTA chelation technique of Sognen (1967) exhibited a weak uptake of glycine to a final distribution ratio of less than 2.0, but the uptake was not sensitive to DNP, raising the possibility that active accumulation was not involved. The calculated distribution ratios which exceeded unity were all small and might be attributed to small errors in estimating intracellular volume. In repeated experiments, Sognen was unable to demonstrate any uptake of 3-O-methyl-glucose (3-OMG) against a concentration gradient. The cell preparation of Huang (1965) produced by the action of lysozyme apparently exhibited limited capability for active uptake of both a sugar (3-OMG) and an amino acid (tyrosine). Fourfold gradients of each were claimed, but 3.3 mM DNP caused only a 44% decrease in uptake, indicating that in reality gradients less than 2-fold had been established. No experiments were performed to examine the characteristics of the uptake system (i.e., Na^+ dependence, K^+ inhibition, ouabain sensitivity) to see if they matched those reported for intact intestine.

In light of the limited information available regarding transport capability of epithelial cells isolated from the small intestine, it seemed imperative that we examine such capability for cells prepared by the hyaluronidase method. In any situation where transport capability of a cell suspension is to be examined, a method is required for rapidly separating the cells from their suspending medium and for rinsing away any solute trapped in the extracellular water of the cell pellet which might otherwise be interpreted as transported material. Two methods are commonly used to accomplish this end: filtration methods with rinsing of the cell pellet on the filter surface and methods involving centrifugation of aliquots taken from the suspension with resuspension and centrifugation required to remove extracellular material trapped in the initial pellet.

Fig. 11

Several reasons led us to consider a filtration method as perhaps the most promising approach. Experience with isolated mitochondria had indicated the rapidity with which a suspension of these organelles can be separated from the medium, and we had used Millipore filtration techniques successfully for monitoring ion transport capability of mitochondria.

Fig. 11. Filter assembly. (a) Millipore filtration apparatus employed for rapid separation of isolated intestinal epithelial cells from the suspending medium. (b) Aliquots of the cell suspension are introduced directly on to filters resting on a sintered glass disc. (c) Clamping devices, such as the one shown, which "sandwich" the filter between two sections of a filter holder are to be avoided for the reasons given in the text. The clamped filter assembly shown here was used to generate the data shown in Table VII.

Filtration and washing of the pellet can typically be accomplished in a few seconds, thereby providing little chance for loss of previously accumulated material. In contrast, centrifugation typically would require a matter of minutes in order to complete both the initial medium separation and the washing procedure. More minutes must pass if several samples are collected before all are centrifuged, and it seemed unlikely that even maintaining the samples chilled on ice would completely prevent diffusional losses of cellular solute. Furthermore, when several samples are handled simultaneously, those collected first of necessity experience a longer lag time between sample collection and processing. Any diffusional loss of material would therefore be expected to be more significant for early than for late samples. Finally, pelleted cell samples require some solubilization or extraction technique in order to release cellular material so an aliquot can be taken for isotope quantitation. These procedures introduce the possibility of quenching or chemiluminescence during the quantitation of isotope by liquid scintillation detection techniques. On the other hand, Millipore filters bearing cell pellets can simply be dried, placed in scintillation vials with counting cocktail, and counted directly. Little mechanical quenching is experienced with the stronger-emitting isotopes (0.15 MeV or greater, e.g., ^{14}C), and literally all of the cells in a given sample can be used rather than an aliquot, as is usually true with centrifugation techniques. Fewer cells and less isotope are consequently required for the filtration method.

Figure 11 shows a photograph of the filtration apparatus which we have used successfully for monitoring transport capability of epithelial cell suspensions. The cell aliquots, which we typically take with a 200-μl micropipette, are introduced onto the surface of a 0.65-μm pore size Millipore filter held on a sintered glass disc (Millipore Filter Corp.). The discs are fitted in rubber stoppers which fit the necks of 250-ml side-arm filtration flasks to which negative pressure can be applied with the aid of a vacuum pump. The flasks are attached to a manifold by short segments of rubber tubing, the air flow through each of which can be controlled by a simple pinch clamp. We find that six to eight flasks connected to a single manifold will operate well if a high-capacity, good-quality vacuum pump is employed. A pump which can move approximately 20 liters of air per minute is appropriate. Even with a pump having the specifications cited, rapid filtration can be accomplished only if just one filtration port is open at a time. Under these conditions, a 200-μl cell sample containing 0.75–1.0 mg cell protein can be filtered in less than 10 s. The cell pellet can be rinsed with 5.0 ml of ice-cold medium which does not contain the radioactive isotope in another 10 s. The port for that flask is then closed and the next one opened in readiness for the next sample. One person working quickly can take samples, filter, wash, and be ready for the next sample within 30 s. If one person samples and another washes the cell pellets and manipulates the apparatus pinch-clamps, samples can be taken at 20-s intervals, but intervals shorter then this are impossible due to the rate of filtration of sample and wash medium.

Using the filtration technique just described, we have been able to characterize the transport capability of isolated intestinal epithelial cells in a variety of ways. Figure 12 shows the change in cellular galactose concentration under various conditions as a function of time of incubation in the presence of 1 mM [^{14}C]galactose (0.15 μCi $^{14}C/\mu$mol). Note that a

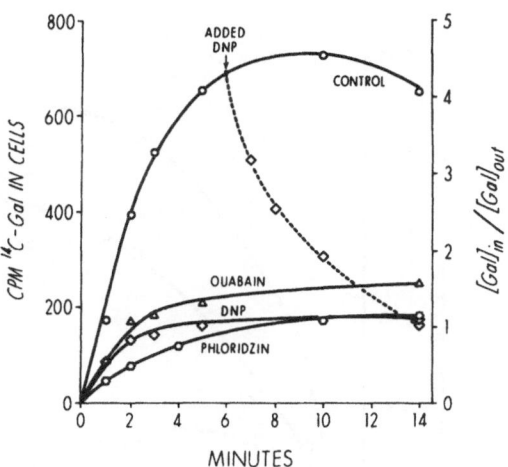

Fig. 12. Accumulation of [^{14}C]-galactose by isolated intestinal epithelial cells, and the effect of several inhibitors. Concentration of each inhibitor used was DNP, 200 μM; ouabain, 125 μM; phloridzin, 200 μM.

rapid entry of galactose occurs under the experimental conditions imposed and that it is very sensitive to the addition of the sugar transport inhibitor phloridzin, the metabolic uncoupler 2,4-dinitrophenol (DNP), and the monovalent cation transport inhibitor ouabain. Each of these agents is known to inhibit active sugar accumulation by intact intestinal tissue, and each by an independent mechanism. Their actions on the isolated cells are a significant clue that the transport capability of the cells has not been destroyed during the isolation procedure or altered in its characteristics. Phloridzin has been shown to be a rather specific competitive inhibitor of the sugar transport system in intact tissue, and is believed to act by binding to the sugar carrier and forming a nonfunctional complex (Caspary et al., 1969). Other investigators have noted that phloridzin can also act as a metabolic inhibitor and interfere with cellular energy production (Newey et al., 1959; Lotspeich and Keller, 1956). However, at the concentration employed for our experiment (200 μM), minimal metabolic effects are noted (Newey et al., 1959). This is further indicated by the fact that cellular ATP levels are maintained (Table VI) and that no inhibitory effect due to phloridzin is noted on the active transport of amino acids, as shown in Fig. 13. In contrast, metabolic inhibitors such as DNP cause a severe decrease in cellular ATP levels (Table VI) and are equally effective on both sugar and amino acid uptake by the cells (Figs. 12 and 13).

Sensitivity of sugar and amino acid uptake by the cells to various metabolic and transport inhibitors does not necessarily imply that the solute has been taken up against a concentration gradient. On the other hand, if a concentration gradient has been established by the cells at the steady-state

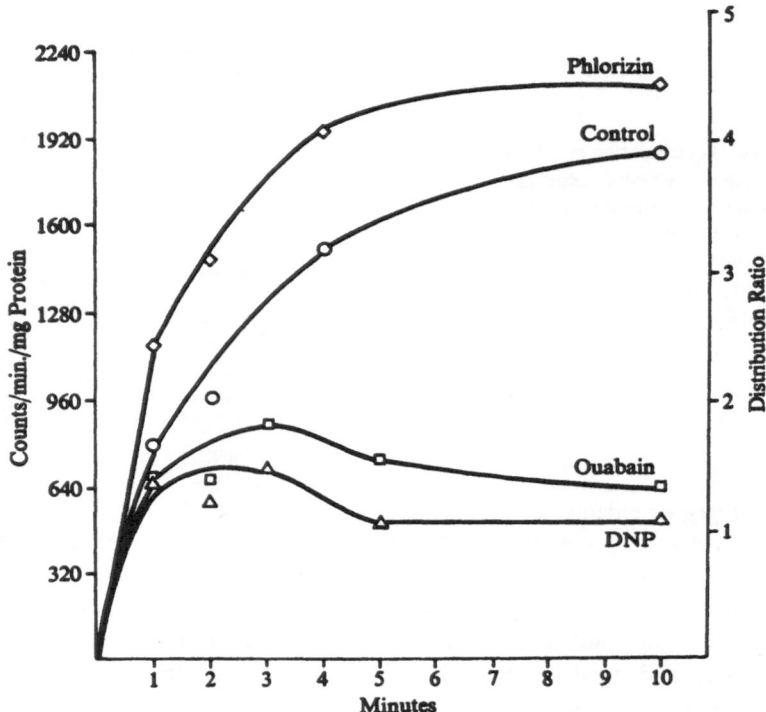

Fig. 13. Effect of phloridzin, ouabain, and DNP on the active accumulation of [^{14}C]valine by isolated intestinal epithelial cells. Inhibitor concentrations were the same as indicated for Fig. 12.

stage of solute uptake, then adding an agent which interrupts cellular energy conservation events should allow the accumulated material to leak back into the medium, moving by diffusion down its concentration gradient. Figures 12 and 14 indicate that rapid efflux of both sugars and amino acids can be induced by adding DNP to a cell population which has reached a steady-state stage of solute uptake. The efflux continues until the cellular content of solute is the same as would have been reached if the inhibitor had been included at the start of the experiment, indicating that this amount of solute represents the point at which intracellular concentration matches that in the suspending medium. We feel that this simple test is of singular importance in establishing whether a given cell population is capable of accumulating a solute against a concentration gradient. It does not depend on an independent determination of intracellular volume, as do most other methods for estimating cellular concentration gradients. The fact that the same steady-state level of accumulation is reached when inhibitor is added either before or after active accumulation indicates that the observed steady state represents a situation in which the cells have loaded to at least a distribution ratio of unity. (Metabolic inhibitors added after active accumulation could not produce active efflux, i.e., extrusion against a concentration gradient.) In contrast, an apparent steady state achieved with inhibitor present from the start might conceivably represent a distribution ratio considerably less than 1.0 if nonactive entry is slow; a gradual approach to equilibrium is difficult to distinguish from a true steady state. If the steady state achieved in the presence of inhibitor indicates a distribution ratio of unity, it is only necessary to take the ratio of intracellular solute in the control case to that in the inhibited steady state in order to calculate the distribution ratio at any point during the experiment. Given the difficulty in accurately estimating intracellular water content (see next section), we feel that this method for determining cellular concentra-

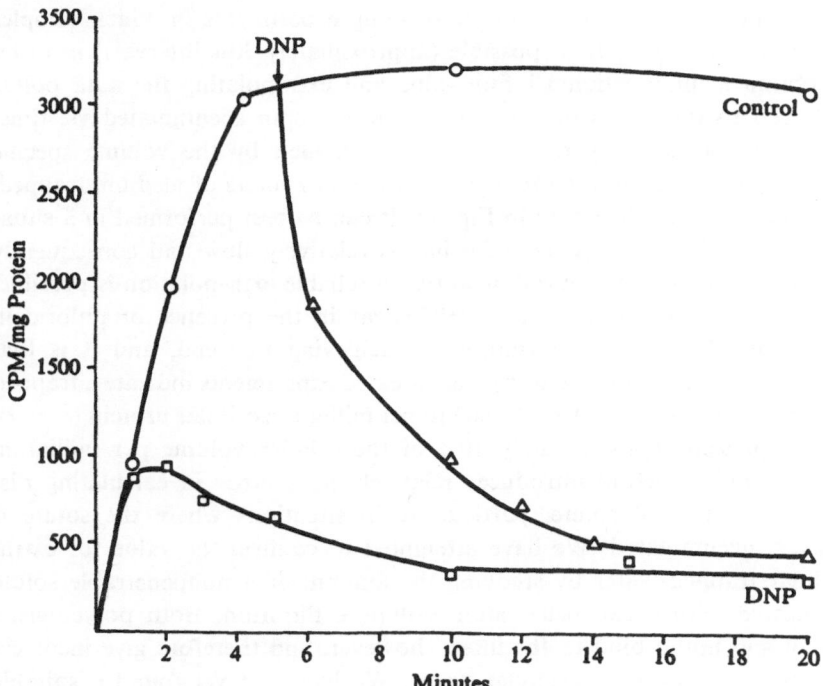

Fig. 14. Efflux of previously accumulated [¹⁴C]valine from isolated intestinal cells induced by the addition of 200 μM DNP.

tion gradients is an underutilized approach which offers the advantages of speed and ease of execution in addition to providing reliable values. For the data shown in Figs. 12 and 14, concentration gradients of approximately 4- and 8-fold were established for 3-*O*-methyl glucose and valine, respectively. Typically, gradients between 4- and 8-fold are established by chick cells prepared by the hyaluronidase method, when extracellular solute concentration is 1 mM.

Two other aspects of the Millipore filtration method need to be briefly mentioned. First, an estimate of the volume of intracellular water in which a given solute distributes can be readily determined simply by dividing the cpm of material at the steady state in fully inhibited cells by the volume specific activity (cpm/μl) of solute in the suspending medium. The volume determined in a particular case can be normalized to a convenient reference parameter such as the amount of cellular protein. In a large number of experiments ($n = 37$), we have found a mean value of 2.3 ± 0.15 μl cell water per milligram cell protein. It is also possible to obtain an estimate of the amount of extracellular water trapped by a cell pellet on the Millipore filter. The estimate is made by performing experiments in which samples are taken as frequently as possible (approximately 30-s intervals) in order to obtain a unidirectional influx value and extrapolating the data points obtained to the *y*-axis of a graph illustrating cpm accumulated vs. time. The cpm indicated by the *y*-intercept is divided by the volume specific activity of the medium in order to calculate microliters of medium trapped. The procedure is illustrated in Fig. 15. It can be best performed in a situation where the rate of entry of solute is relatively slow and consequently linear for a reasonable duration so that a reliable extrapolation is possible. We have found that uptake of [^{14}C]sugar in the presence of phloridzin represents the best circumstances for achieving this end, and it is that situation which is illustrated. Again, repeated experiments indicate a trapped extracellular volume of 0.25 ± 0.03 μl per milligram cellular protein ($n = 9$). This represents approximately 10% of the cellular volume per milligram protein and therefore introduces relatively little error in calculating distribution ratios of solute, particularly in situations where the solute is actively accumulated. We have attempted to confirm the value for extracellular trapped water by studying the amount of a nonpenetrable solute associated with a cell pellet after Millipore filtration. Both polyethylene glycol and inulin bind to the filters, however, and therefore give incorrect values for trapped extracellular water. We have not yet found a suitable extracellular marker which can be employed for making this measurement when filtration is used as the means of separating cells from medium.

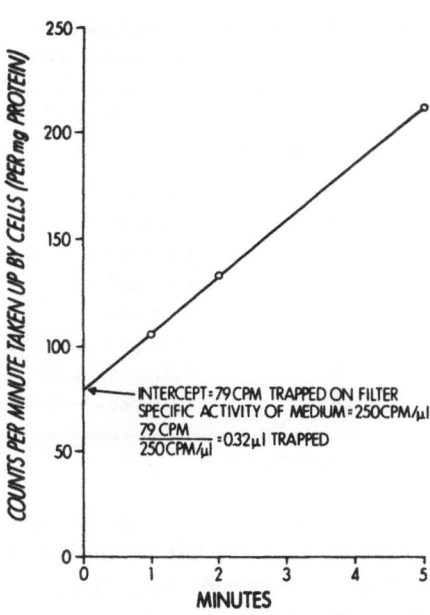

Fig. 15. Method used to calculate the volume of extracellular fluid trapped by the cell pellet after Millipore filtration separation of cells from suspending medium. Details are described in the text.

Before leaving the topic of Millipore filtration, the question of binding of solutes to the filters must be considered further. Reiser and Christiansen (1971) have concluded that transport studies involving separation of intestinal cells by Millipore filtration techniques cannot be conducted successfully due to large and variable binding of solute (amino acid) to the filter. Personal communication with a relatively large number of investigators indicated that they had experienced similar difficulty. In trying to assess the reason for these reports in light of our own success with the filtration method, we were led to consider any differences in experimental technique among the various laboratories. We soon learned that one common denominator to the technique used by those groups experiencing difficulty was clamping of the filter between two pieces of filter apparatus, in contrast to our situation in which it merely rests on a filter support (see Fig. 11). It seemed possible that material trapped at the filter–apparatus junction during the initial filtration might move laterally through filter capillaries into the clamped filter edge, and therefore be rather effectively shielded from ready access to the medium used to wash the exposed filter area. We tested this possibility simply by clamping a filter chimney to the filter holder as shown in Fig. 11c, filtering a volume of medium containing a total of 110,000 cpm [^{14}C]valine, washing with 10 ml of nonradioactive medium, and comparing the counts retained by the filter to those retained

Table VII. Radioactivity Retained by Millipore Filter After Filtering Medium Containing 110,000 cpm [^{14}C]Valine

Filter No.	Filtration geometry	Cpm retained − Bkg[a]
1	Unclamped	5
2	Unclamped	8
3	Unclamped	2
4	Clamped	239
5	Clamped	359
6	Clamped, whole filter	537
6	Clamped, center of filter	36
6	Clamped outer 3 mm	489

[a] Background 25 cpm.

by a similar filter in which the "chimney" had not been employed. Data are shown in Table VII. Note that 50–100 times more radioactivity remains on the "sandwiched" filter than on the free filter and that the variability of counts remaining is very large when the clamping device is used. Furthermore, if only the central area of the clamped filter is cut out and counted, it too has a low count rate, while the clamped area of the filter has the bulk of the total radioactivity (see Table VII). In fact, the outer 3 mm of the filter retains 90% of the counts trapped, although it represents only 42% of the total filter area. This information indicates that the isotope detected represents material trapped in the filter–apparatus junctional area rather than solute binding to the filter. We therefore have concluded that most difficulties attributed to variable filter binding of amino acids or sugars in reality represent mechanical trapping of solute at points where the filter is "sandwiched" in the apparatus. In a situation where a cell suspension is filtered instead of medium without cells, it seems logical that mechanical trapping of isotope would be even more troublesome. In the case of Reiser and Christiansens's (1971) experiments, a Swinny-type filter holder was employed and the medium was forced through the filter with the aid of pressure applied with a syringe. In light of the relatively large amounts of protein used, high applied pressure was necessary, which might force still more material into the junctional area where it would be poorly accessible to wash liquid. Extremely large and variable corrections would be necessary and were reported. The applied pressure might also damage the cell population and further account for the relatively poor solute gradients observed when the clamped filter apparatus was employed.

4.2. Centrifugation Techniques

Because of the difficulties experienced by Reiser and Christiansen (1971) in employing Millipore filtration for the separation of cells from incubating medium, they adopted a centrifugal separation technique. Although the solute gradients which they observed were not large (approximately 2-fold or less), they did report that successive washing of the cell pellet with ice-cold saline did not lead to loss of radioactive solute taken up during the prior incubation interval. This report indicated that our earlier reservations regarding the suitability of centrifugation as a separation technique might have been unwarranted. Furthermore, if samples could be taken and stored for a short interval prior to centrifugation, it seemed likely that more samples could be obtained in a given interval than when sampling was limited by the rate of sample filtration. Rapid short-term sampling could be quite useful for certain kinetic experiments in which unidirectional influx rates must be measured. We therefore decided to further evaluate centrifugation as a tool for cell separation from incubation media.

If the technique is to be successful, it is important that transmembrane solute fluxes be prevented as soon as possible after the sample aliquot is taken. Rather than simply chilling the sample aliquot and hoping that the rate of cooling was rapid enough to immediately prevent solute flux, we decided to dilute the sample into a relatively large volume of previously chilled medium. This technique has the double advantage of more immediate chilling of the cells as well as dilution of the radioactivity in the sample so that any trapped extracellular fluid after the first centrifugation will carry a less significant portion of the total isotope associated with the pellet. With the aid of [^{14}C]polyethyleneglycol, we have found that when a 200-μl cell sample (0.5–1.0 mg protein) is diluted into 2.0 ml of ice-cold medium, only 15 μl of extracellular fluid is trapped in the pellet after centrifugation of 1 min at 5000g. Considering the dilution factor involved, this is equivalent to only 1.4 μl of the original incubation medium. When the pellet is resuspended in 2.0 ml of chilled medium and recentrifuged, the new pellet will carry extracellular fluid equivalent to only 0.01 μl of the original incubation medium, an amount which can safely be ignored. In order to ascertain the amount of isotope associated with the final pellet, representing accumulated solute, we simply extract with 5% trichloroacetic acid, centrifuge down denatured protein, and count an aliquot of the supernatant by liquid scintillation spectrometry. A tissue solubilizer is added to the counting cocktail in order to allow the toluene to accommodate the aqueous sample.

We have found that BBS-3 (Beckman Instruments, Inc.) is a particularly suitable solubilizer which does not chemiluminesce and produces practically no quenching. Best success has been attained if the aqueous sample is fully solubilized in relatively concentrated BBS-3 and toluene before the counting cocktail is brought to its final volume with toluene containing the scintillation fluors.

Using the technique described above, we have been able to confirm all of the cellular transport characteristics established by the Millipore filtration procedure. Figure 16 shows a comparison of data obtained for 3-OMG uptake by each of the two methods applied to cells from a single preparation. Multiple experiments have indicated no significant difference in the rate or extent of solute accumulation detected by the two methods, despite the fact that cells handled by the centrifugation method are in contact with their suspending medium a much longer period of time. Apparently, rapid and continued chilling of the cells is fully effective in limiting trans-membrane solute fluxes in either an inward or an outward direction.

Furthermore, as we had predicted, the centrifugation method is some-what more amenable to obtaining samples in short time intervals. It is no trouble for one person to obtain samples every 15 s, and, with practice, intervals as short as 10 s can be accommodated. Figures 17 and 18 illustrate

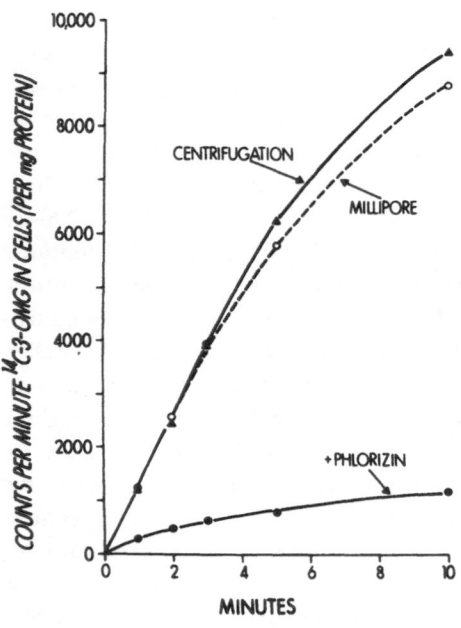

Fig. 16. Comparison of ^{14}C-labeled 3-O-methylglucose (3-omb) accumulated by isolated intestinal epithelial cells as determined by Millipore filtration or centrifugation techniques for separating cells from their suspending medium.

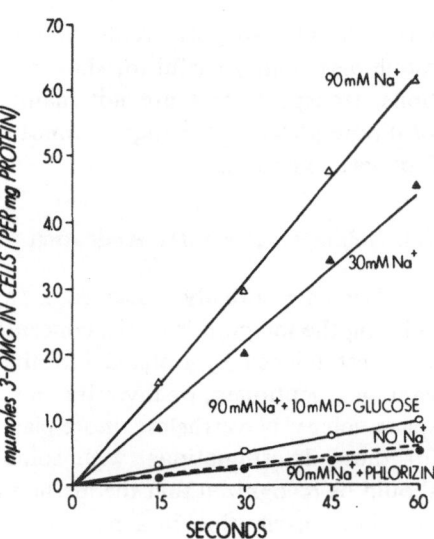

Fig. 17. Effect of 200 μM phloridzin, 10 mM D-glucose, or lack of Na+ on the unidirectional influx of [14C]-labeled 3-O-methylglucose (3-omg) into isolated intestinal epithelial cells. Cells were separated from the incubation medium by the centrifugation technique described in the text.

short-term experiments in which the uptake of either 1 mM 3-OMG or 1 mM valine was monitored. Note the excellent degree of linearity attained, indicating that true unidirectional influx, uncomplicated by efflux, is being observed. Unidirectional flux rates are essential for adequate kinetic charac-

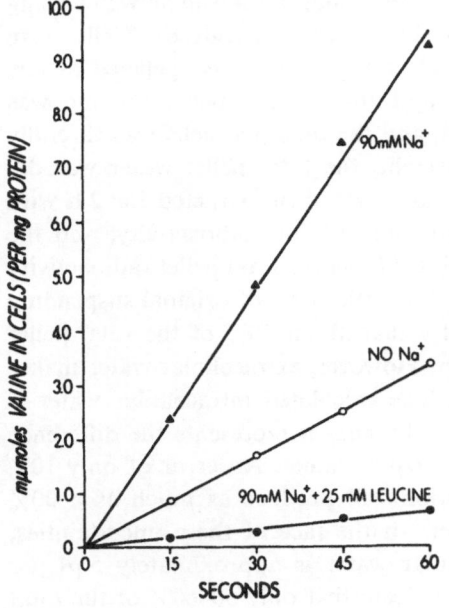

Fig. 18. Effect of Na+ concentration and 25 mM leucine on the unidirectional influx of [14C]valine into isolated intestinal epithelial cells. Cells were separated from the suspending medium by centrifugation.

terization of transport events, and we have found the centrifugation technique particularly useful for short-term kinetic studies. Under many conditions, transport rates are not maintained linear for even 1 min, and the Millipore filtration technique cannot be satisfactorily employed for accurate flux determinations.

4.3. Solute Gradient Determination

We have already discussed a rapid and convenient method for calculating the magnitude of the concentration gradients for sugars and amino acids established by isolated intestinal epithelial cells. In light of the advantages mentioned, we favor it as a routine tool for calculating intracellular space values. Nevertheless, more classical techniques can also be employed for these determinations, with some interesting results. For instance, it should be recognized that the method described earlier based on the volume of solute distribution in a metabolically inhibited cell population yields a space for only that part of the total cell water that is penetrated by the solute employed. Recognizing that not all cellular organelles are necessarily equally permeable to various solutes, and that some bound or otherwise inaccessible water may exist within the cell, it is logical to expect that total cell water may be somewhat different from the penetrable cell water.

We evaluated this possibility by determining intracellular water using the membrane-impermeable marker [^{14}C]polyethyleneglycol. Cells were suspended in the presence of [^{14}C]polyethyleneglycol and pelleted at low speed in a finely calibrated McNaught tube. Total pellet volume was recorded, the supernate was removed, and the tube plus cells was carefully weighed. After drying to constant weight, the total pellet water was determined by difference. The pellet residue was then extracted for 2 h with 1 N HNO$_3$, an aliquot was taken for determining radioactivity, and the trapped extracellular water was calculated from the total pellet radioactivity and the cpm of [^{14}C]polyethyleneglycol per microliter of original suspending medium. This value typically indicates that about 50% of the total pellet volume consists of extracellular water. However, extracellular water makes up about 80% of total pellet water. The calculated intracellular water is therefore subject to considerable error because it represents the difference between two large and approximately equal values. An error of only 10% in the extracellular water determination will produce as much as a 30% error in calculated intracellular water. In the face of these uncertainties, our best estimate for total intracellular water is approximately 5 μl per milligram cell protein, which would indicate that only 50–60% of the total cell water is penetrated by the free sugar or amino acid molecules in the presence of a transport inhibitor.

The latter value can be confirmed if a second cell sample is pelleted in the presence of 100 μM phloridzin plus [^{14}C] 3-OMG. The water penetrated by the sugar under these conditions represents extracellular water *plus* accessible cell water. If the polyethyleneglycol-penetrable water (exclusively extracellular) from the first experiment is subtracted, the remainder is 2.5–3.0 μl per milligram cell protein, which is in good agreement with the value obtained by filtration experiments in which the corrections for trapped extracellular water are much more modest. A similar value is attained for cell water accessible to amino acid in the presence of DNP. While there is considerable error involved in these experiments, we feel it is significant that the values calculated for solute-penetrable water are so similar whether determined directly (Figs. 12 and 13) or by difference, as described above. It is surprising that only 60% of the total cell water is accessible, but unclear whether poor accessibility represents lack of penetration of certain organelle membranes (nucleus, mitochondria, etc.) or exclusion due to an ordered array of water molecules at certain cellular loci. In all likelihood, a combination of factors is involved.

5. TRANSPORT CHARACTERISTICS

5.1. Sugar and Amino Acid Transport

Having established the fact that intestinal cells isolated by the hyaluronidase method have the capability for active transport of sugars and amino acids, the next question to consider is whether the characteristics of transport are the same as have been established for various preparations of intact tissue. Using the procedures already described, we have been able, in fact, to prove that the transport characteristics of the isolated cells are very similar to those recognized as typical of intestinal tissue. We have described these observations in detail in several other publications (Kimmich, 1970b; Tucker and Kimmich, 1973; Kimmich and Randles, 1973a,b) and will mention only the primary points here. Figure 17 shows a set of experiments in which unidirectional influx of [^{14}C] 3-OMG was examined in several different situations. It indicates that a very high proportion of the flux is inhibited by removing Na$^+$ from the incubation medium, by including the sugar transport inhibitor, phloridzin, or by including glucose, which shares the sugar carrier with 3-OMG. We have calculated initial flux rates from

Table VIII. Comparison of Unidirectional Fluxes for Amino Acids and 3-OMG Established for Intact Intestinal Tissue or Isolated Intestinal Cells Incubated Under Various Conditions

Substrate	Preparation	Condition	Unidirectional influx[a]	Percent control
Alanine (5 mM)	Intact tissue[b]	Control	2.2	—
		No Na+	0.6	27
		+20 mM valine	0.38	17
Valine (1 mM)	Cells	Control	9.1	—
		No Na+	3.4	37
		+25 mM leucine	0.66	7.3
3-OMG (20 mM)	Intact tissue[b]	Control	2.53	—
		No Na+	0.31	12.2
		+Phloridzin	0.22	8.7
		+20 mM glucose	0.25	9.65
3-OMG (1 mM)	Cells	Control	6.10	—
		No Na+	0.60	9.8
		+phloridzin	0.52	8.5
		+10 mM glucose	0.96	11.4

[a] Values are $\mu mol/cm^2 \cdot$ h for intact tissue preparation and $\mu mol/min \cdot$ mg protein for isolated cells.
[b] Data for intact tissue are taken or calculated from Goldner et al. (1969).

Fig. 17 and tabulated them along with unidirectional fluxes observed by Goldner et al. (1969, 1972) for segments of intact rabbit tissue (Table VIII). Note that in each case a very high percentage of the observed influx was dependent on the presence of Na+ in the incubating medium and could be inhibited with phloridzin or by including a second actively transported sugar. These characteristics have long been recognized as typical of intestinal solute transport systems and generally have been ascribed to entry systems restricted to the brush border membrane. Neither phloridzin sensitivity nor Na+ dependence has ever been demonstrated for influx of sugar across the serosal membrane of epithelial cells in intact tissue. We therefore believe that the transport phenomena observed with the intact cells represent primarily brush border transport systems and can be used to characterize the nature of those systems. This is particularly true for sugar transport, where brush border carrier-mediated entry can be operationally defined as phloridzin-sensitive entry.

Na$^+$-dependent entry can also be readily demonstrated for various neutral amino acids, as is shown in Fig. 18 for the case of valine. In this case, phloridzin is of course noninhibitory, but other neutral amino acids such as leucine produce a striking inhibition, indicating competition for a common carrier system. Again a high percentage of total influx is Na$^+$ dependent, as has been noted in experiments performed with intact tissue, in which only brush border entry rates were examined (Schultz *et al.*, 1967) (see Table VIII). An inhibitor which is specific for brush border amino acid uptake would be valuable for further clarifying this point, but unfortunately none has been described. High concentrations of leucine inhibit more than 90% of the unidirectional influx of valine, as shown in Fig. 18, but this might represent competition for valine carriers at cell surfaces other than those associated with the brush border. However, when we consider the interaction between sugar and amino acid transport systems we will present further evidence indicating that brush border amino acid entry represents a large part of the total cellular uptake.

Aside from the stimulatory role which Na$^+$ plays in enhancing unidirectional solute fluxes, we have also demonstrated that Na$^+$ is required for active transport capability. A representative experiment is shown in Fig. 19, in which 3-OMG uptake was monitored at several different Na$^+$ concentrations. The degree of accumulation achieved in the absence of Na$^+$ is the same as that observed in the presence of phloridzin and represents a final distribution ratio of unity. Note that as little as 20 mM Na$^+$ allows the cells to establish approximately a 4-fold concentration gradient of 3-OMG,

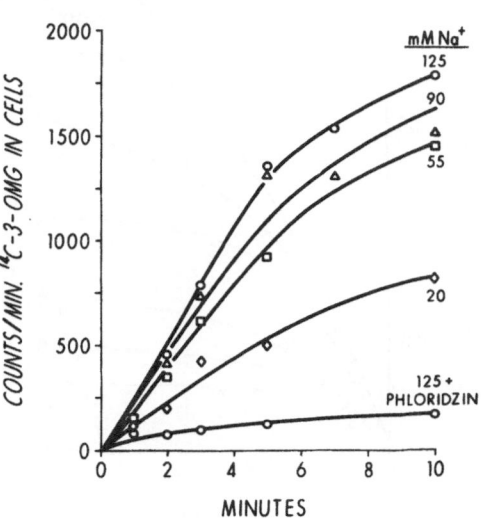

Fig. 19. Effect of medium Na$^+$ concentration on the accumulation of [^{14}C] 3-OMG by isolated intestinal epithelial cells.

and near-maximal gradients are established when Na^+ is raised to 80 mM. Kinetic analysis indicates that half-maximal influx rates are supported by approximately 20 mM Na^+, in agreement with values determined for intact tissue.

Another characteristic of transport as assessed in intact intestinal tissue is sensitivity to elevated K^+ concentrations. The isolated epithelial cells exhibit similar sensitivity, as shown in Fig. 20, providing still further evidence that tissue transport characteristics have not been altered during the cell isolation procedure. When medium Na^+ concentration is 20 mM, as much as 90% of the active solute accumulation can be inhibited by 90 mM K^+. At higher Na^+ concentrations, K^+ is less inhibitory, suggesting that K^+ sensitivity is a result of competition for the Na^+ binding site of the solute carrier, as has been concluded by Crane *et al.* (1965).

Before concluding this brief survey of the general transport characteristics of the isolated intestinal epithelial cells, mention should be made of the action of certain metabolic inhibitors. In light of the very rapid glycolytic activity exhibited by the cells, it seemed possible that glycolysis might be largely responsible for supporting energy-dependent transport events. If so, a part of the active transport capability might not be sensitive to inhibitors of the respiratory chain, which would block energy production due to oxidative phosphorylation but still allow glycolysis to occur. In contrast to the prediction, we find that active transport is extremely sensitive to low concentrations of rotenone, even in the presence of mannose, which does

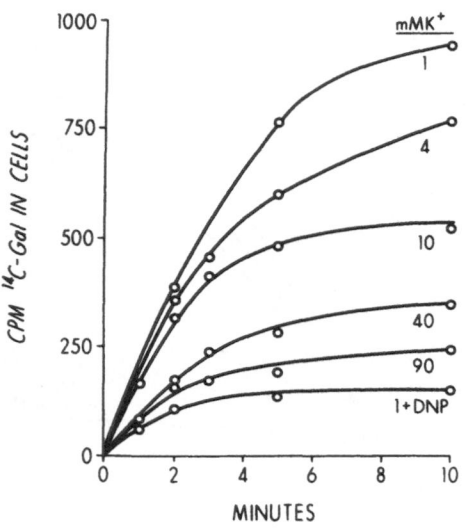

Fig. 20. Effect of elevated K^+ concentration on the accumulation of [^{14}C] galactose by isolated intestinal epithelial cells. The medium Na^+ concentration was 20 mM in each case. Osmolarity was maintained constant in each case by providing the appropriate amount of mannitol.

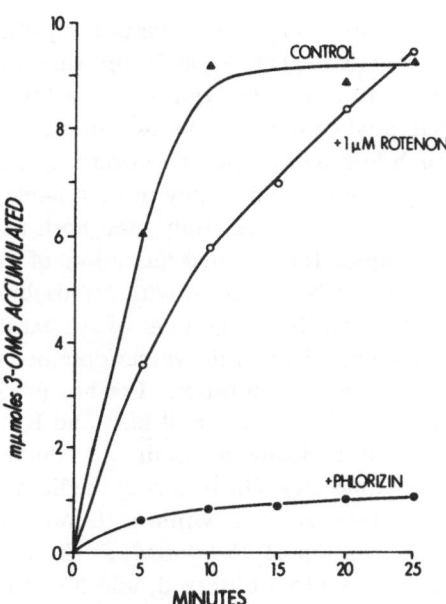

Fig. 21. Effect of a low concentration of rotenone on the active accumulation of [^{14}C] 3-OMG by isolated intestinal epithelial cells. The incubation medium was supplemented with 10 mM mannose.

not interfere with the entry system for 3-OMG, but which can be readily glycolyzed (see Fig. 21). Half-maximal inhibition occurs at a rotenone concentration of about 2 µM. In the absence of a metabolizable sugar, only 1 µM rotenone exerts half-maximal inhibition, indicating that glycolytic energy production can afford a degree of protection against the inhibitor but that oxidative metabolism is essential for fully maintaining energy-dependent events. A particularly intriguing aspect of the effect of low concentrations of rotenone is the fact that while lower rates of uptake are observed the final steady-state gradients are the same or even greater than in those cases where rotenone is absent (see Fig. 21). Apparently, rotenone or enhanced glycolytic activity induces a decrease in both influx and efflux rates of solute. It seems possible that these data may indicate a degree of control of solute efflux at serosal cell boundaries induced by changes in glycolytic activity. This possibility is being further evaluated.

5.2. Monovalent Ion Transport

In addition to the solute transport systems, it is possible to readily demonstrate active transport of the monovalent cations Na$^+$ and K$^+$ by the isolated cells. Transport of K$^+$ is the more readily documented since it represents active accumulation against a concentration gradient rather than active extrusion as is the case with Na^+. Furthermore, medium K^+ concentrations are typically very low relative to Na^+ concentrations, so that any contamination of samples with extracellular medium entails a much less significant correction factor for K^+. Values obtained for cellular K^+ are thus considerably more reliable than those calculated for Na^+.

We have successfully used both Millipore filtration and centrifugation techniques for the determination of cellular K^+. In either case, the cell pellets can be extracted with 5% trichloroacetic acid for 1 h, then properly diluted and the K^+ content of the extract determined by flame photometry with comparison of the values obtained to those observed with K^+ standards of known concentration. Freshly prepared cells which have been preincubated in the absence of Na^+ and K^+ apparently lose a large part of their K^+ and typically maintain a cellular concentration of only 25–30 mM (see Table IX). On restoring medium Na^+ and K^+, however, they readily accumulate K^+ and within a 10-min interval can restore cellular levels to a concentration of 75–90 mM, as shown in Table IX. Values higher than 90 mM are seldom observed, and 90 mM is somewhat lower than the values reported for other types of cells, which indicates a degree of leakiness of the cell membrane to K^+. The exceptionally high turnover of energy associated with monovalent ion transport (see Section 3.2) is in all likelihood a reflection of the high membrane permeability toward these ions. Whether a high K^+ permeability is a result of alterations in membrane characteristics

Table IX. K^+ Content of Intestinal Cells Preincubated in Na^+-Free Medium Containing 0 or 80 mM K^+ After Transfer to Medium Containing 60 mM Na^+

Time elapsed after transfer of cells to medium with 60 mM Na^+ and 20 mM K^+ (min)	nmol K^+/mg cell protein: cells preincubated at	
	0 mM K^+	80 mM K^+
0	33	85
1	28	88
3	45	88
5	50	—
8	63	88
10	75	69

Data taken from Kimmich and Randles (1973a).

during preparation of the cells or whether it reflects a situation physiologically characteristic of intestinal tissue is a matter for debate. Schultz *et al.* (1966) have reported cell K^+ concentrations near 140 mM for strips of epithelium prepared from rabbit intestine. On the other hand, Armstrong *et al.* (1970) have found approximately 85 mM K^+ in the epithelial cells of intact bullfrog intestine. The latter value has recently been confirmed by direct measurement with the aid of ion-selective microelectrodes (Lee and Armstrong, 1972). Further work is required in this area with intact chick intestine before an answer can be fully given.

Determination of cellular Na^+ concentrations is slightly more problematic. Even the slight degree of contamination of the cell pellet on a Millipore filter with extracellular medium represents a considerable fraction of the total pellet Na^+, particularly when medium Na^+ concentrations are high. Furthermore, the filters themselves contain a variable amount of Na^+ as purchased which may represent as much of the ion as that retained by the cell pellets. These two factors render Millipore filtration unusable as a means of separating cells from medium for cellular Na^+ determinations. Instead, it is necessary to introduce cell samples to a chilled Na^+-free medium for centrifugal separation. Resuspension and centrifugation in a few milliliters of fresh ice-cold Na^+-free medium are effective for reducing extracellular Na^+ to a low level even when medium Na^+ concentrations are originally 100 mM or greater. The cell pellet can then be extracted and assayed for Na^+ as described earlier using flame photometry techniques. Using this procedure, we have found cellular Na^+ concentrations of 35 mM after a short incubation in our standard medium containing 80 mM Na^+, 80 mM mannitol, 6 mM $KHPO_4$, 20 mM tris-Cl, 1 mM $MgCl_2$, 1 mM $CaCl_2$, and 1 mg/ml bovine serum albumin. Adding ouabain caused cellular Na^+ to rise to approximately 55 mM within 15 min. The control value for average cellular Na^+ concentration is in very good agreement with the value of 29 mM determined by Armstrong *et al.* (1970) for bullfrog intestine incubated under comparable conditions. These determinations do not discriminate between true intracellular Na^+ and Na^+ bound to the exterior surface of the cell. However, isotopes can be used effectively to demonstrate that transmembrane fluxes of material are actually involved in the changes observed for cellular Na^+. The best procedure for demonstrating such fluxes is to incubate the cell population at 4°C in the presence of $^{22}Na^+$ and to suddenly introduce the preloaded cell population into a Na^+-free medium from which samples can be taken for cell separation by Millipore filtration. If the dilution of the extracellular Na^+ concentration is great enough so that the intracellular Na^+ concentration is initially greater than the extra-

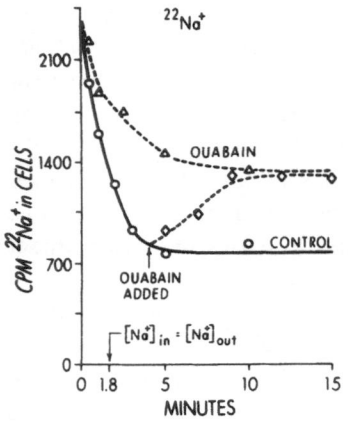

Fig. 22. Extrusion of ^{22}Na$^+$ by intestinal cells pre-loaded with ^{22}Na$^+$ (at 80 mM Na$^+$) and incubated at 20 mM Na$^+$. Ouabain (125 μM) was either present from the start or added to a control experiment after 4 min incubation. From Kimmich (1970*b*).

cellular, then efflux of isotope would be expected by diffusion and due to activity of the outwardly directed Na$^+$ pump. Loss of ^{22}Na$^+$ would be expected until the cells establish a normally directed Na$^+$ gradient, at which point a steady-state level of ^{22}Na$^+$ should be maintained. If the cells are introduced into medium containing an inhibitor of the Na$^+$ pump such as ouabain, a slower rate and lesser extent of ^{22}Na$^+$ efflux are expected as only diffusional forces operate. Figure 22 shows that exactly this pattern of events is observed if the cells are preincubated at 80 mM Na$^+$ and introduced to an environment containing 20 mM Na$^+$. Furthermore, if ouabain is added after the control cells have extruded Na$^+$ to a steady state, there is an immediate reentry of ^{22}Na$^+$ which continues until a new steady state is reached that corresponds to what is observed when ouabain is present from the beginning. The magnitude of the concentration gradient which is established for Na$^+$ does not appear to be large on the basis of the difference in steady-state levels maintained by inhibited and control cells. On the other hand, it is logical to expect considerable binding of Na$^+$ to intracellular anionic constituents so that the cellular concentration of Na$^+$ does not accurately reflect the activity of Na$^+$. If diffusional efflux in the presence of ouabain continues until equilibrium is reached, then the steady state represents the point at which cellular Na$^+$ activity matches that in the suspending medium. The difference between the inhibited steady state and the calculated distribution ratio of unity should provide an indication of bound Na$^+$. For the case illustrated in Fig. 22 (extracellular Na$^+$ = 20 mM), it can be estimated that approximately one-third of the total cellular Na$^+$ is bound when ouabain is present. If the same amount of bound Na$^+$ is subtracted from the control case, it is possible to estimate that approximately a 3-fold concentration gradient for Na^+ has been established under the particular incubation conditions chosen. While these estimates are only approximate, it is encouraging to note that recent work by Lee and Armstrong (1972) with Na^+-sensitive microelectrodes indicates that about half of the total cellular Na^+ in bullfrog intestine is present in bound form.

5.3. Interaction between Transport Systems

A number of investigators (Newey and Smyth, 1964; Alvarado, 1966; Chez *et al.*, 1966; Semenza, 1971) have noted that there is a mutually inhibitory interaction between transport systems for sugars and amino acids in intestinal tissue from a wide variety of animal species. Such interaction has been noted so consistently with intact tissue preparations that it seemed important to establish whether the phenomenon could also be demonstrated with isolated cells as a further aspect of evaluating their functional integrity. Again, our work in this area has been described in detail (Kimmich and Randles, 1973b) and will be cited here only to the extent that it bears on concepts to be treated in the next section.

Figures 23 and 24 show that there is indeed a very pronounced interaction between sugar and amino acid transport systems which can be demon-

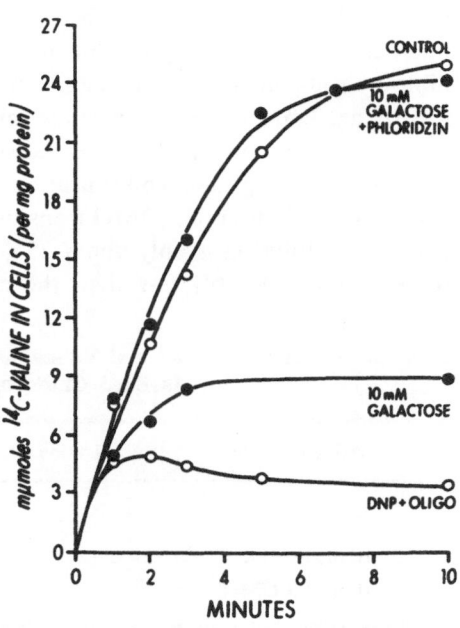

Fig. 23. Effect of 10 mM galactose on the accumulation of 1 mM [¹⁴C]valine by isolated intestinal cells. Note that inclusion of 200 µM phloridzin completely relieves the inhibitory effect induced by galactose.

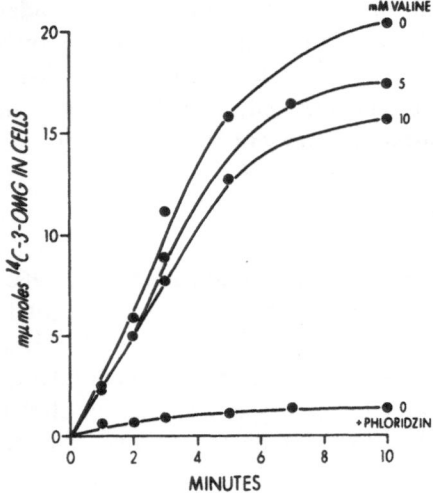

Fig. 24. Effect of 10 mM valine on the accumulation of 1 mM [¹⁴C] 3-OMG by isolated intestinal cells.

strated using cells isolated by the hyaluronidase method. As much as 75% of the active accumulation of 1 mM valine can be inhibited by including 10 mM 3-OMG in the incubation medium. Higher sugar concentrations exert very little further inhibitory effect, and 200 μM phloridzin completely relieves the inhibition. These facts suggest that the sugar must be able to interact with its entry system in order to act as an inhibitor and that inhibition is maximal when the sugar carrier is saturated. Sugar concentrations less than 10 mM exert intermediate inhibitory effects in proportion to the expected degree of saturation of the sugar carrier.

Conversely, graded concentrations of valine cause successively greater degrees of inhibition of 3-OMG transport, as shown in Fig. 24. The maximal degree of inhibition is only about 25% of the active 3-OMG uptake and is therefore considerably less than the effect exerted by the sugar. If both

Table X. Effect of 3-OMG and Valine on the Accumulation of 1 mM Lysine by Isolated Chick Intestinal Cells

Added solute	Lysine accumulated (nmol/mg protein in 15 min)
None	8.14
10 mM valine	7.84
10 mM 3-OMG	6.70

3-OMG and valine are studied for the effect they produce on the entry of a third solute which utilizes still a different carrier, a very similar relationship is observed. As shown in Table X, 3-OMG inhibits 15% of active lysine accumulation, while no inhibition is produced by valine. The significance of these observations will be considered in Section 7.

6. USE OF ISOLATED CELLS FOR EVALUATING CONCEPTS RELATING TO THE MECHANISM OF INTESTINAL TRANSPORT

In light of the similarity of the functional capability of intestinal epithelial cells isolated by the hyaluronidase procedure to that of intact tissue preparations, it seemed likely that the cells might be useful for gaining further mechanistic insight into the nature of the Na^+-dependent transport systems. As mentioned at the outset of this chapter, a widely recognized important test of the ion gradient hypothesis involves imposing a transmembrane Na^+ gradient of reversed polarity and examining the direction of net solute flux induced in a cell population which has been preloaded with solute to a distribution ratio of unity. If the ion gradient hypothesis is correct, an initial outward extrusion of solute would be expected in response to the reversed Na^+ gradient. As the cells lose Na^+ due to diffusion and to the action of the Na^+ extrusion mechanism, the capability for active extrusion of solute should be lost. Finally, only when the cells have extruded sufficient Na^+ to allow them to begin to reestablish a Na^+ gradient of normal polarity would active accumulation of solute against a concentration gradient be expected.

We have already demonstrated that a reversed Na^+ gradient can be transiently imposed on isolated intestinal epithelial cells by a procedure involving preincubation of the cell population in a high Na^+ medium and sudden dilution into a Na^+-free medium (see Fig. 22). If the steady-state cellular Na^+ concentration in the presence of inhibitor can be taken as the point at which cellular activity of Na^+ matches the activity of Na^+ in the suspending medium, inspection of Fig. 22 indicates that noninhibited cells require approximately 2 min to reduce cellular Na^+ to that level. We therefore have a situation where an imposed reversed Na^+ gradient can be maintained for almost 2 min, allowing that amount of time to examine the rate and direction of net flux for a previously equilibrated solute. Figures 25 and 26 show two experiments in which the effect of a reversed Na^+ gradient on net flux of 3-OMG and valine was examined. In each case, cells were preincubated with 80 mM Na^+ (plus $^{22}Na^+$) and

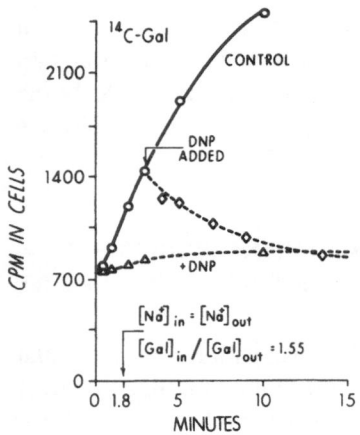

Fig. 25. Accumulation of [¹⁴C]galactose by isolated intestinal cells in which a reversed Na^+ gradient had been imposed. Cells were "loaded" in medium containing 80 mM Na^+ and 1.25 mM galactose, at 0°C, and incubated at 37°C in medium containing 1.25 mM galactose and 20 mM Na^+. These data were obtained with the same cell preparation used to monitor the $^{22}Na^+$ fluxes shown in Fig. 22. After 3 min of incubation, 200 μM DNP was added to a control preparation.

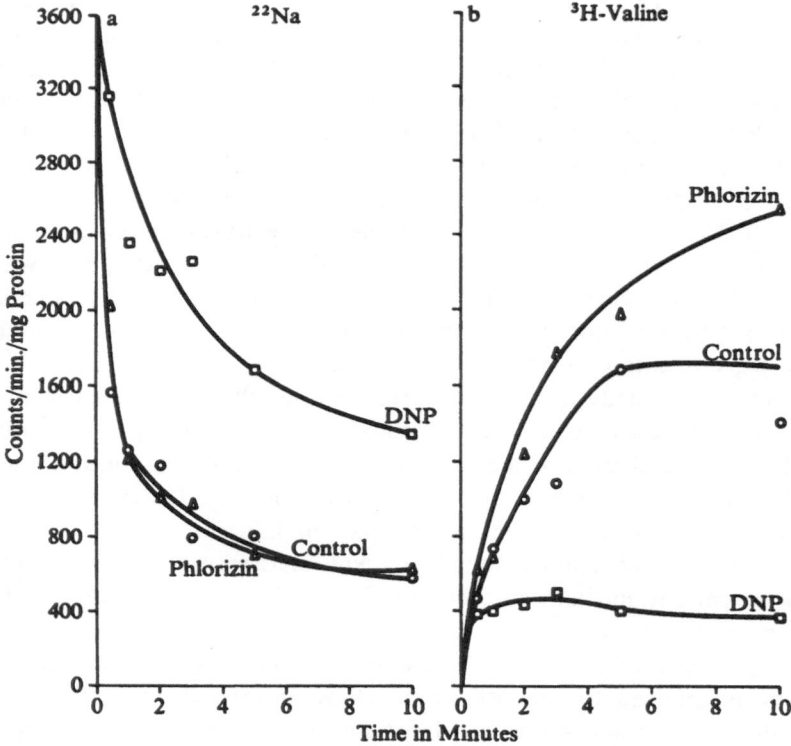

Fig. 26. Extrusion of $^{22}Na^+$ and accumulation of [³H]valine by isolated intestinal cells preincubated with 80 mM $^{22}Na^+$ at 0°C and incubated at 37°C with 1 mM [³H]valine and 20 mM $^{22}Na^+$.

diluted into Na^+-free medium so that the final extracellular Na^+ concentration became 20 mM. Prior to the dilution, a 1-min incubation at 37°C with the radioactive solute at 1 mM was allowed in order to facilitate equilibration of solute between cellular and medium water. Control experiments indicate that a 1-min incubation is adequate for establishing a solute distribution ratio of unity. The Na^+-free diluent medium contained solute at the same concentration and specific activity as during the preincubation.

In contrast to predictions derived from the ion gradient hypothesis, no initial efflux of solute from the cells was observed. Instead, an immediate and rapid solute influx occurred which persisted until a new steady state of solute distribution was observed after about 10 min of incubation. The influx rate during the first 2 min of incubation when the Na^+ gradient was reversed in polarity was as rapid as during any subsequent 2-min interval when a normally directed Na^+ gradient had began to be reestablished. The fact that the early flux represents a true active accumulation of solute can be readily demonstrated. If a metabolic inhibitor such as DNP is added shortly after the experiment is initiated, there is an immediate efflux of solute which proceeds to exactly the same level as achieved when the inhibitor is included from the start (see Fig. 25). Apparently, solute has been accumulated against a concentration gradient even at this early point in time (while the Na^+ gradient is reversed) and the inhibitor induces diffusional loss until an equilibrium distribution is reached.

All of the above facts are difficult to reconcile with a conceptual mechanistic model for solute transport which depends solely on the polarity and magnitude of transmembrane Na^+ gradients as a determinant of the polarity and magnitude of solute distribution ratios. We have also tested the possibility that the cellular K^+ gradient might play a role as determinant of transport capability, but again with no positive correlation. These experiments are reported in detail elsewhere (Kimmich and Randles, 1973b); they were similar in format to those described above, and will not be detailed here.

While experiments in which cellular Na^+ concentrations are manipulated are much more reliably performed with an isolated cell population than with intact tissue preparations, some serious pitfalls exist which must be taken into account. For instance, it must be recognized that cellular Na^+ concentrations measured either directly or indirectly with the aid of isotope represent average cell concentrations and do not necessarily reflect the concentration sensed by the solute carrier systems at the interior surface of the plasma membrane. It is possible that microenvironments exist near the membrane surface in which the Na^+ concentration is significantly lower than that calculated as average cellular Na^+ concentration. It must also be recognized that the likelihood exists that not all of the cells in a given population are equally viable, as indicated by the dye exclusion experiments. It is possible that the results obtained above, in which reversed Na^+ gradients were imposed, reflect different capabilities of two cell populations with differing degrees of viability. The indication of Na^+ loading and diffusional loss of Na^+ in the presence of ouabain could represent a relatively nonviable population of cells with little solute transport capability, while the observed sugar accumulation might represent function of a more viable population which was maintaining a Na^+ gradient of normal polarity.

These possibilities are difficult to evaluate fully, but a consideration of the basis of interaction between transport systems for sugars and amino acids suggests a reasonable approach. In terms consistent with the ion gradient hypothesis, interaction is thought to be a result of partial dissociation of cellular Na^+ gradients during transport of one solute (due to cotransport of Na^+), leaving less energy inherent in the Na^+ gradient to support accumulation of a second solute. If this concept is accurate, one might logically expect a high degree of correlation between the rate of transport of a solute and its ability to inhibit accumulation of a second solute. That is, dissipation of the cellular Na^+ gradient should depend on the coentry rate of Na^+ with solute, which is in turn dependent on solute entry rate.

In light of the foregoing comments and the data for interaction between transport systems for valine and 3-OMG shown in Figs. 23 and 24, we expected that 10 mM 3-OMG might be transported at a much higher rate than 10 mM valine. Instead, we found valine to be transported approximately 60% faster than 3-OMG. Even after correcting for expected differences in the stoichiometry of coupling between Na^+ and solute entry, it can be calculated that valine would be expected to promote Na^+ coentry more than 50% faster than 3-OMG, as shown in Table XI. These facts relate to a population of cells which are maintaining active transport capability, and are not subject to the ambiguity associated with those experiments in which reversed Na^+ gradients had been imposed, where viable cells might have been responsible for sugar fluxes and nonviable cells for the observed Na^+ fluxes. If Na^+ delivery on the sugar carrier to the cell interior (or microenvironment) is able to inhibit 75% of active valine accumulation, it is unlikely that an even greater rate of Na^+ delivery on the valine carrier into the same cell population (or microenvironment) would be only one-third as effective in inhibiting 3-OMG accumulation.

Table XI. Comparison of Relative Inhibitory Effectiveness of 10 mM Valine or 10 mM 3-OMG with Predicted Degree of Discharge of Cellular Na$^+$ Gradient[a]

Substrate	Entry rate (nmol/min · mg protein)			Na$^+$ coupling coefficient	Na$^+$ entry rate[b]	Induced Na$^+$ entry Val/3-OMG	%I_{Val}/%I_{3-OMG}[c]
	Total	Passive	Carrier mediated				
Valine	19	3.0	16	0.8	12.8		
3-OMG	12.5	3.0	9.5	1.0	9.5	1.35	0.33

[a] Values reported for rabbit ileum incubated with 80 mM Na$^+$ (Curran et al., 1967; Goldner et al., 1969).

[b] Na$^+$ entry rate = carrier-mediated substrate entry rate × coupling coefficient.

[c] %I_{Val} = percent inhibition produced by 10 mM valine on the uptake of 1 mM 3-OMG. %I_{3-OMG} = percent inhibition produced by 10 mM 3-OMG on the uptake of 1 mM valine. %I_{Val}/%I_{3-OMG} should agree well with Na$^+$ entry Val/3-OMG if interaction between transport systems is due to competition for energy inherent in the cellular Na$^+$ gradient.

Low concentrations of ouabain, which partially dissipate cellular Na^+ gradients, inhibit 3-OMG and valine accumulation to identical degrees, indicating equal sensitivity of the carriers to changes in cellular Na^+. Note that this approach is based on a consideration of an active accumulation event which would require a functional cell population, and therefore is not complicated by the possible presence of nonviable cells with no active transport capability. We thus are faced with a second experimental approach utilizing isolated intestinal epithelial cells which yields data which are difficult to reconcile with concepts implicit in the ion gradient hypothesis.

7. PROBLEMS RELATED TO ASCERTAINING POLARITY OF FUNCTION FOR ISOLATED INTESTINAL EPITHELIAL CELLS

We have briefly summarized data which we feel bear on concepts related to the mechanism of Na^+-dependent transport systems. Some of the experimental approaches described are particularly well suited to studies involving the use of isolated cells, especially those in which attempts are made to manipulate cellular Na^+ concentrations. The problem of a mixed population of viable and nonviable cell is difficult to deal with, although the same difficulty would apply to comparable experiments performed with intact tissue. In the latter case, the problem is compounded by the presence of a wide variety of cell types and relatively significant tissue compartments of poorly accessible extracellular space. When tissue Na^+ concentration is manipulated, it is impossible to know to what extent changes in total tissue Na^+ apply to the columnar epithelial cells.

A more frequent objection to work with isolated cells relates to the fact that it is difficult to discern their polarity of transport capability. With intact tissue, the use of transport chambers allows only one surface of the gut wall to be exposed to solute, so that brush border transport systems can be characterized in a manner uncomplicated by transport events occurring at other cell surfaces. In working with isolated cell populations, both brush border and lateral-serosal cell boundaries are exposed and solute fluxes at all surfaces can contribute to observed transport rates. Concern has been expressed that characterization of brush border transport might be impossible by methods involving the use of isolated cells.

While the concern is legitimate, we feel that a number of observations indicate that solute accumulation by the isolated cells represents primarily brush border transport capability. A particularly strong case can be developed for the case of sugar transport by making use of phloridzin. Phloridzin has been studied extensively and has been shown to be a potent inhibitor of epithelial cell sugar accumulation when added at the mucosal boundary of intact tissue, but not if added at the serosal surface (Newey et al., 1959; Kinter et al., 1965). Autoradiographic studies indicate that phloridzin does not readily penetrate the epithelial cell membrane and exhibits localized binding at the brush border boundary of the cell (Stirling, 1967). Alvarado and Crane (1962) have shown that the agent exhibits competitive inhibition kinetics on sugar transport, but again found little evidence of penetration of the cell membrane. Finally, while Kinter et al. (1965) concluded that there might be a weak transport system for sugars at the serosal cell boundary, they could find no evidence of phloridzin sensitivity. All of these facts suggest that phloridzin acts rather specifically on active transport systems localized in the brush border with no detectable effect on sugar transfer events associated with the lateral-serosal cell boundary.

In light of this information, the high degree of sensitivity to phloridzin of sugar uptake by the isolated epithelial cells indicates that the accumulation is primarily mediated by brush border transport systems. Indeed, 92% of the initial entry rate is inhibited by phloridzin, as shown in Fig. 17. Goldner et al. (1969) found that 91% of the unidirectional flux of 3-OMG across the brush border surface of rabbit intestine was inhibited by phloridzin. Nearly equal dependence on extracellular Na^+ was noted for sugar uptake by both the isolated cells and intact tissue exposed at only the mucosal surface. Lack of Na^+ inhibited 90% of the unidirectional influx observed with isolated cells (Fig. 17) and 85% of that observed with the intact tissue. We believe that such marked sensitivities to phloridzin and Na^+ deprivation indicate an entry route for the intact cells which is primarily mediated by the brush border localized transport system.

In the case of amino acid uptake, we do not have the advantage of a specific inhibitor for those entry routes localized in the brush border membrane. It is encouraging, however, that approximately 75% of active valine accumulation can be inhibited by 3-OMG in a manner completely relieved by phloridzin. Again, if the action of phloridzin can be related to its effect at the brush border boundary, this observation may indicate that a high proportion of amino acid entry occurs at the mucosal pole of the cell. Furthermore, we have shown that methionine inhibits 92% of the initial undirectional flux of valine. Although methionine might have been expected to inhibit carrier-mediated valine entry at any cell surface, Kinter et al. (1965) have shown that serosal entry of valine was completely insensitive to the presence of methionine. Finally, in those situations where unidirectional fluxes across serosal boundaries have been calculated for sugars and amino acids from data obtained with intact tissue, the rate is typically less than 15% of the rate measured for brush border fluxes (Schultz *et al.*, 1967; Munck and Schultz, 1969; Alvarado, 1968; Goldner *et al.*, 1969). The extracellular Na^+ concentration has no significant effect on the serosal influx rate. It seems likely therefore that serosal fluxes contribute only modestly to the total fluxes observed with the isolated cells. While it is possible that limited active transport occurs at the lateral-serosal cell boundaries, any such activity must be Na^+ dependent and ouabain inhibited (Figs. 13 and 18), and therefore itself probably due to a mechanism similar to that postulated for the brush border transport systems. Those experiments designed to probe aspects of the transport mechanism would therefore not be unduly complicated by the presence of low-capacity serosally located systems of similar fundamental design in terms of energetic coupling.

8. CONCLUSION

The final chapter can not yet be written. We feel that we have provided important evidence indicating that intestinal epithelial cells can be successfully isolated with retention of their metabolic and functional capability. We believe that such preparations offer significant advantages over intact tissue preparations for executing certain experiments designated to evaluate concepts related to the mechanism of the Na^+-dependent transport systems. While some of those experiments have provided data which are difficult to reconcile with currently established mechanistic concepts, it is possible that problems inherent in the use of randomly oriented isolated cells can account for some or all of the conceptual discrepancies derived from data from very different systems. At present, we believe that many of the basic objections to the use of isolated cells are unfounded, or can be circumvented. We are continuing our attempts at evaluating the use of isolated cells for transport studies and only with further experimentation and development of methods will complete answers be achieved. It is already clear that data obtained with isolated cells cannot be fully interpreted without supporting information obtained from various types of intact tissue preparations for estimating fluxes associated with particular cell boundaries. At the same time, we believe the fullest understanding of intestinal transport systems will be achieved through the use of well-defined homogeneous cell preparations in tandem with established classical techniques utilizing intact tissue. We expect increasing acceptance of the use of isolated cells by the scientific community and increasing application to various areas of transport and metabolic investigation.

ACKNOWLEDGMENTS

The work described was supported partially by grants from the U.S. Public Health Service, No. AM 15365 and No. AM 70166, and in part under contract with the U.S. Atomic Energy Commission at the University of Rochester Atomic Energy Project and has been assigned Report No. U.R.-3490-589.

9. REFERENCES

Abbott, W. O., and Miller, T. G., 1936, Intubation studies of the human small intestine. III. A technique for the collection of pure intestinal secretion and for study of intestinal absorption, *Am. Med. Assoc.* **106**:16.

Agar, W. T., Hird, F. J. R., and Sidhu, G. S., 1954, The uptake of amino acids by the intestine, *Biochim. Biophys. Acta* **14**:80.

Agar, W. T., Hird, F. J. R., and Sidhu, G. S., 1956, The absorption, transfer, and uptake of amino acids by intestinal tissue, *Biochim. Biophys. Acta* **22**:21.

Alvarado, F., 1966, Transport of sugars and amino acids in the intestine: Evidence for a common carrier, *Science* **151**:1010.

Alvarado, F., 1968, Amino acid transport in hamster small intestine: Site of inhibition by D-galactose, *Nature* **219**:276

Alvarado, F., and Crane, R. K., 1962, Phlorizin as a competitive inhibitor of the active transport of sugars by hamster small intestine, *in vitro*, *Biochim. Biophys. Acta* **56**:170.

Armstrong, W. McD., Musselman, D. L., and Reitzug, H. C., 1970, Sodium, potassium, and water content of isolated bullfrog small intestinal epithelia, *Am. J. Physiol.* **219**:1023.

Barrett, E. G., 1974, Ultrastructural and transport properties of isolated intestinal epithelial cells, Ph.D. Thesis, University of Rochester.

Barrett, E. G., and Coleman, J. R., 1973, Sodium and potassium content of single cells: Effects of metabolic and structural changes, in: *Proceedings of the 8th National Conference on Electron Probe Analysis*, New Orleans, p. 60.

Bergstrom, S., Blomstrand, R., and Borgstrom, B., 1954, Route of absorption and distribution of oleic acid and triolein in the rat, *Biochem. J.* **58**:600.

Biggs, M. W., Friedman, M., and Byers, S. O., 1951, Intestinal lymphatic transport of absorbed cholesterol, *Proc. Soc. Exp. Biol. Med.* **78**:641.

Bihler, I., and Crane, R. K., 1962, Studies on the mechanism of intestinal absorption of sugars: The influence of several cations and anions on the active transport of sugars, *in vitro*, by various preparations of hamster small intestine, *Biochim. Biophys. Acta* **59**:78.

Blomstrand, R., Borgstrom, B., and Dahlback, O., 1959, Extent of total hydrolysis of dietary glycerides during digestion and absorption in the human, *Proc. Soc. Exp. Biol. Med.* **102**:204.

Bloom, B., Chaikoff, I. L., Reinhardt, W. O., Entenman, C., and Dauben, W. G., 1950, The quantitative significance of the lymphatic pathway in transport of absorbed fatty acids, *J. Biol. Chem.* **184**:1.

Bollman, J. L., Cain, J. C., and Grindlay, J. H., 1948, Techniques for the collection of lymph from the liver, small intestine, or thoracic duct of the rat, *J. Lab. Clin. Med.* **33**:1349.

Caspary, W. F., Stevenson, N. R., and Crane, R. K., 1969, Evidence for an intermediate step in carrier-mediated sugar translocation across the brush border membrane of hamster small intestine, *Biochim. Biophys. Acta* **193**:168.

Chaikoff, I. L., Bloom, B., Siperstein, M. L., Kiyasu, J. Y., Reinhardt, W. O., Dauben, W. G., and Eastham, J. F., 1952, C^{14}-Cholesterol. I. Lymphatic transport of absorbed cholesterol-4-C^{14}, *J. Biol. Chem.* **194**:407.

Chapman, A. G., Fall, L., and Atkinson, D. E., 1971, Adenylate energy charge in *Escherichia coli* during growth and starvation, *J. Bacteriol.* **108**:1072.

Chez, R. A., Schultz, S. G., and Curran, P. F., 1966, Effect of sugars on transport of alanine in intestine, *Science* **53**:1012.

Clark, B., and Porteus, J. W., 1965, The isolation and properties of epithelial cell "ghosts" from rat small intestine, *Biochem. J.* **96**:539.

Cori, C. F., 1925, The fate of sugar in the animal body. I. The rate of absorption of hexoses and pentose from the intestinal tract, *J. Biol. Chem.* **66**:691.

Crane, R. K., and Mandelstam, P., 1960, The active transport of sugars by various preparations of hamster intestine, *Biochim. Biophys. Acta* **45**:460.

Crane, R. K., Miller, D., and Bihler, I., 1960, The restrictions on possible mechanisms of intestinal active transport of sugars, in: *Symposium on Membrane Transport and Metabolism* (A. Kotýk and A. Kleinzeller, eds.), pp. 439–449, Academic Press, New York.

Crane, R. K., Forstner, G., and Eicholz, A., 1965, Studies on the mechanism of the intestinal absorption of sugars. X. An effect of Na^+ concentration on the apparent Michaelis constants for intestinal sugar transport *in vitro*, *Biochim. Biophys. Acta* **109**:467.

Cummins, A. J., and Jussila, R., 1955, Comparison of glucose absorption rates in the upper and lower human small intestine, *Gastroenterology* **29**:982.

Curran, P. F., Schultz, S. G., Chez, R. A., and Fuisz, R. E., 1967, Kinetic relations of the Na^+–amino acid interaction at the mucosal border of intestine, *J. Gen. Physiol.* **50**:1261.

Darlington, W. A., and Quastel, J. H., 1953, Absorption of sugars from isolated surviving intestine, *Arch. Biochem.* **43**:194.

Dent, C. E., and Schilling, J. A., 1949, Studies on the absorption of protein: The amino acid pattern in the portal blood, *Biochem. J.* **44**:318.

Dickens, F., and Weil-Malherbe, H., 1941, Metabolism of normal and tumor tissue. 19. The metabolism of intestinal mucus membrane, *Biochem. J.* **35**:7.

Fisher, R. B., and Parsons, D. S., 1949, A preparation of surviving rat small intestine for the study of absorption, *J. Physiol. (London)* **110**:36.

Fisher, R. B., and Parsons, D. S., 1953a, Glucose movements across the wall of the rat small intestine, *J. Physiol. (London)* **119**:210.

Fisher, R. B., and Parsons, D. S., 1953*b*, Galactose absorption from the surviving small intestine of the rat, *J. Physiol. (London)* **119**:224.

Fullerton, P. M., and Parsons, D. S., 1956, The absorption of sugars and water from rat intestine *in vivo, Quart. J. Exp. Physiol.* **41**:387.

Goldner, A. M., Schultz, S. G., and Curran, P. F., 1969, Sodium and sugar fluxes across the mucosal border of rabbit ileum, *J. Gen. Physiol.* **53**:362.

Goldner, A. M., Hajjar, J. J., and Curran, P. F., 1972, Effects of inhibitors on 3-*O*-methylglucose transport in rabbit ileum, *J. Membr. Biol.* **10**:267.

Gornall, A. G., Bardawill, C. S., and David, M. M., 1949, Determination of serum protein by means of the biuret reaction, *J. Biol. Chem.* **177**:751.

Harrer, D. S., Stern, B. K., and Reilly, R. W., 1964, Removal and dissociation of epithelial cells from the rodent gastrointestinal tract, *Nature* **203**:319.

Harrison, D. D., and Webster, H. L., 1969, The preparation of isolated intestinal crypt cells, *Exp. Cell Res.* **55**:257.

Huang, K. C., 1965, Uptake of L-tyrosine and 3-*O*-methylglucose by isolated intestinal epithelial cells, *Life Sci.* **4**:1201.

Iemhoff, W. G. J., Van Den Berg, J. W. O., De Pyper, A. M., and Hulsmann, W. C., 1970, Metabolic aspects of isolated cells from rat small intestinal epithelium, *Biochim. Biophys. Acta* **215**:229.

Jacobs, F. A., and Luper, M., 1957, Intestinal absorption by perfusion *in situ, J. Appl. Physiol.* **11**:136.

Kimmich, G. A., 1970*a*, Preparation and properties of mucosal epithelial cells isolated from small intestine of the chicken, *Biochemistry* **9**:3659.

Kimmich, G. A., 1970*b*, Active sugar accumulation by isolated intestinal epithelial cells: A new model for sodium dependent metabolite transport, *Biochemistry* **9**:3669.

Kimmich, G. A., 1973, Coupling between Na^+ and sugar transport in small intestine, *Biochim. Biophys. Acta* **300**:31.

Kimmich, G. A., and Randles, J., 1973*a*, Effect of K^+ and K^+ gradients on accumulation of sugars by isolated intestinal epithelial cells, *J. Membr. Biol.* **12**:23.

Kimmich, G. A., and Randles, J., 1973*b*, Interaction between Na^+-dependent transport systems for sugars and amino acids: Evidence against a role for the sodium gradient, *J. Membr. Biol.* **12**:47.

Kinter, W. B., Wilson, T. H., and Mullen, D. A., 1965, Autoradiographic study of sugar and amino acid absorption by everted sacs of hamster intestine, *J. Cell Biol.* **25**:19.

Kiyasu, J. Y., and Chaikoff, I. L., 1957, On the manner of transport of absorbed fructose, *J. Biol. Chem.* **224**:935.

Kiyasu, J. Y., Katz, J., and Chaikoff, I. L., 1956, Nature of the C^{14}-compounds recovered in portal plasma after enteral administration of C^{14}-glucose, *Biochim. Biophys. Acta* **21**:286.

Lee, C. O., and Armstrong, W. Mc. D., 1972, Activities of sodium and potassium ions in epithelial cells of small intestine, *Science* **175**:1261.

Levenson, S. M., Rosen, H., and Upjohn, H. Z., 1959, Nature and appearance of protein digestion in upper mesenteric blood, *Proc. Soc. Exp. Biol. Med.* **101**:178.

London, E. S., 1929, Experimental fistulae of blood vessels, *Harvey Lect.* **23**:208.

Lotspeich, W. D., and Keller, D. M., 1956, A study of some effects of phlorizin on the metabolism of kidney tissue *in vitro, J. Biol. Chem.* **222**:843.

Matthews, D. M., and Smyth, D. H., 1954, The intestinal absorption of amino acid enantiomorphs, *J. Physiol. (London)* **126**:96.

Miller, T. G., and Abbott, W. O., 1934, Intestinal intubation; a practical technique, *Am. J. Med. Sci.* **187**:595.

Munck, B. G., and Schultz, S. G., 1969, Interaction between leucine and lysine transport in rabbit ileum, *Biochim. Biophys. Acta* **183**:192.

Newey, H., and Smyth, D. H., 1964, Effects of sugars on intestinal transfer of amino acids, *Nature* **202**:400.

Newey, H., Parsons, B. J., and Smyth, D. H., 1959, The site of action of phlorizin in inhibiting intestinal absorption of glucose, *J. Physiol.* **148**:83.

Ohnell, R., 1939, The artificially perfused mammalian intestine as a useful preparation for studying intestinal absorption, *J. Cell Comp. Physiol.* **13**:155.

Perris, A. D., 1965, Isolation of the epithelial cells of the rat small intestine, *Can. J. Biochem.* **44**:687.

Porteus, J. W., and Clark, B., 1965, The isolation and characterization of subcellular components of the epithelial cells of rabbit small intestine, *Biochem. J.* **96**:159.

Reid, E. W., 1901, Transport of fluid by certain epithelia, *J. Physiol.* (*London*) **26**:436.

Reiser, S., and Christiansen, P., 1971, The properties of the preferential uptake of L-leucine by isolated intestinal epithelial cells, *Biochim. Biophys. Acta* **225**:123.

Schultz, S. G., and Curran, P. F., 1970, Coupled transport of sodium and organic solutes, *Physiol. Rev.* **50**:637.

Schultz, S. G., Fuisz, R. E., and Curran, P. F., 1966, Amino acid and sugar transport in rabbit ileum, *J. Gen. Physiol.* **49**:849.

Schultz, S. G., Curran, P. F., Chez, R. A., and Fuisz, R. E., 1967, Alanine and sodium fluxes across mucosal border of rabbit ileum, *J. Gen. Physiol.* **50**:1241.

Semenza, G., 1971, On the mechanism of mutual inhibition among sodium-dependent transport systems in the small intestine: A hypothesis, *Biochim. Biophys. Acta* **241**:637.

Shay, H., Gershon-Cohen, J., Fels, S. S., and Munro, F. L., 1940, The fate of ingested glucose solution of various concentrations at different levels of the small intestine, *Am. J. Digest. Dis.* **7**:456.

Sheff, M. F., and Smyth, D. H., 1955, An apparatus for the study of *in vivo* intestinal absorption in the rat, *J. Physiol.* (*London*) **128**:67P.

Shoemaker, W. C., Yanof, H. M., Turk, L. N., and Wilson, T. H., 1963, Glucose and fructose absorption in unanesthetized dogs, *Gastroenterology* **44**:654.

Sjostrand, F. S., 1968, A simple and rapid method to prepare dispersions of columnar epithelial cells from the rat intestine, *J. Ultrastruct. Res.* **22**:424.

Sognen, E., 1967, A method for the preparation of suspensions of intestinal mucosal cells by means of calcium chelation, *Acta Vet. Scand.* **8**:76.

Sols, A., and Ponz, F., 1947, A new method for the study of intestinal absorption, *Rev. Espan. Fisiol.* **3**:207.

Stern, B. K., 1966, Some biochemical properties of suspensions of intestinal epithelial cells, *Gastroenterology* **51**:855.

Stern, B. K., and Jensen, W. E., 1966, Active transport of glucose by suspensions of isolated rat intestinal epithelial cells, *Nature* **209**:789.

Stern, B. K., and Reilly, R. W., 1965, Some characteristics of the respiratory metabolism of suspensions of rat intestinal epithelial cells, *Nature* **205**:563.

Stirling, C. E., 1967, High-resolution radioautography of phlorizin-^{3}H in rings of hamster intestine, *J. Cell Biol.* **35**:605.

Thiry, L., 1864, Über eine neue Methode, den Dunndarm zu isolieren, *Sitzungsber. Akad. Wiss. Wien Math.-Naturwiss. Kl. Abt. I* **50**:77.

Tucker, A. M., and Kimmich, G. A., 1973, Characteristics of amino acid accumulation by isolated intestinal epithelial cells, *J. Membr. Biol.* **12**:1.

Van Slyke, D. D., and Meyer, G. M., 1912, The amino acid nitrogen of the blood: Preliminary experiments on protein assimilation, *J. Biol. Chem.* **12**:399.

Vella, L., 1888, Neues Verfahren zur Gewinnung reinen Darmsaftes und Feststellung seiner physiologischen Eigenschaften, *Untersuch. Naturl. Menach Thiere* **13**:40.

von Mering, J., 1877, Ueber die Abzugswege des Zuckers aus der Darmhohe, *Arch. Anat. Physiol.*, 379.

Webster, H. L., and Harrison, D. D., 1969, Enzymic activities during the transformation of crypt to columnar intestinal cells, *Exp. Cell Res.* **56**:245.

Wiggans, D. S., and Johnstone, J. M., 1959, The absorption of peptides, *Biochim. Biophys. Acta* **32**:69.

Wilson, T. H., and Vincent, T. N., 1955, Absorption of sugars *in vitro* by the intestine of the golden hamster, *J. Biol. Chem.* **216**:851.

Wilson, T. H., and Wiseman, G., 1954, The use of sacs of everted small intestine for the study of the transference of substances from the mucosal to the serosal surface, *J. Physiol. (London)* **123**:116.

Wiseman, G., 1953, Absorption of amino acids using an *in vitro* technique, *J. Physiol. (London)* **120**:63.

Wu, R., and Racker, E., 1959, Regulatory mechanisms in carbohydrate metabolism: Limiting factors in glycolysis of ascites tumor cells, *J. Biol. Chem.* **234**:1029.

Chapter 3

Use of Isolated Membrane Vesicles in Transport Studies

JOY HOCHSTADT, DENNIS C. QUINLAN,
RICHARD L. RADER, CHIEN-CHUNG LI,
and DIANA DOWD

Worcester Foundation for Experimental Biology
Shrewsbury, Massachusetts

1. INTRODUCTION

Isolated membrane vesicles represent one of the simplest systems in which the transport processes remain intact and have therefore provided considerable information on the structure, function, and regulation of transport systems. Although whole cells may also be used for transport studies, it is often difficult to separate transport from subsequent events in intracellular intermediary metabolism. One of the important advantages offered by membrane vesicles is the opportunity to investigate their transport functions apart from other cellular activities. At the same time, the isolated transport systems of membrane vesicles are retained in a functional form and their relationship to other membrane components can also be studied.

Bacterial cells as a source for membrane vesicles (Kaback and Stadtman, 1966) have the additional advantages of population homogeneity, ability to grow on defined media and to respond to environmental manipulations, and the availability of a variety of metabolic mutants.

Consequently, work with isolated bacterial membrane vesicles has led to the elucidation of the transport mechanisms for sugars (Kaback, 1968; Barnes and Kaback, 1970), amino acids (Kaback and Milner, 1970), purine bases (Hochstadt-Ozer and Stadtman, 1971*b*), purine nucleosides (Hochstadt-Ozer, 1972), pyrimidine bases (Hochstadt-Ozer and Cashel, 1972), pyrimidine nucleosides (Hochstadt, 1974), and cations (Bhattacharyya *et al.*, 1971; Lombardi *et al.*, 1973; Barnes, 1974).

In addition to the study of transport, bacterial membrane vesicles have been used to investigate the role of the membrane in cell division (Shapiro *et al.*, 1970), oxidative phosphorylation (Klein and Boyer, 1972), electron transport (Short *et al.*, 1974; Kaback, 1970; Barnes and Kaback, 1970), nucleotide binding (Weissbach *et al.*, 1969), and phospholipid biosynthesis (Kaback and Stadtman, 1968).

Vesicles derived from mammalian surface membranes were isolated and a number of their biochemical and biophysical properties studied almost 10 years ago (Wallach *et al.*, 1966; Wallach and Kamat, 1966*a*; Wallach and Zahler, 1966). These experiments involved plasma membrane vesicles isolated from noncloned, heterozygous cells grown *in vivo*. Although subsequent biochemical studies have utilized membrane vesicles isolated from mammalian cell lines grown in culture (Wu *et al.*, 1969; Meezan *et al.*, 1969), in order to maximize the genetic and metabolic advantages inherent in the bacterial systems, almost all transport studies with mammalian membrane vesicles were with vesicles derived from cells grown *in vivo* (Carter and Martin, 1969; Carter *et al.*, 1972). The use of established mammalian tissue culture lines as a source of membrane vesicles in order to study transport is therefore quite recent (Hochstadt, 1974).

Transport studies employing established mammalian cell lines in which uptake into isolated membrane vesicles is emphasized may be particularly important for investigating the recently proposed possibility that nutrient transport is a key process in regulation of cell growth (Holley, 1972; Cunningham and Pardee, 1969; Quinlan and Hochstadt, 1974). The homogeneity of such cell populations would be necessary for study of various growth states (e.g., cell cycle phases or quiescence of a confluent culture), while the isolated vesicle system would be necessary for determination of the way the membrane itself participated in growth control.

Although many studies have been devoted to nutrient transport in mammalian systems, most have employed intact cells. Thus transport mechanisms (facilitated diffusion vs. classical, active transport vs. group translocation) have not yet been resolved in a definitive manner since metabolism carried out by membrane-bound enzymes and metabolism subsequent to transport cannot be separated in an intact cellular system (e.g., see Schuster and Hare, 1971; Plagemann and Erbe, 1972). Also, the energetics of transport in mammalian cells have not been elucidated. In contrast, the use of membrane vesicles has led to a significantly greater understanding of energy coupling in bacterial substrate transport systems (see Kaback, 1972).

Our approach to mammalian transport has been to employ established cell lines and their derived membrane vesicles for the study of (1) nucleoside transport in membrane vesicles derived from homozygous cell lines grown on completely defined, serum-free medium (Li and Hochstadt, 1975), (2) nucleoside and base transport by membrane vesicles isolated from quiescent and actively growing 3T3 cells and SV40-transformed 3T3 cells (Quinlan and Hochstadt, 1974, 1975a, b), (3) nucleoside transport in membrane vesicles isolated from normal and polyoma-transformed baby hamster kidney fibroblasts (Dowd and Hochstadt, in preparation), and (4) the role of phosphoribosyltransferase in hypoxanthine transport and the role of the Na^+/K^+-activated Mg^+-ATPase in providing energy for transport (Quinlan et al., in prep.) using membrane vesicles from thioguanine- and ouabain-resistant 3T3 cell lines, respectively.

The brief introduction has attempted to emphasize the advantages of using isolated membrane vesicles, as opposed to whole cells, in order to study a function unique to membranes, transport, both in bacteria and in mammalian cells. The remainder of this chapter presents the background, rationale, and methodology of preparation of membrane vesicles from bacterial and mammalian cells used in this laboratory. However, our aim is not only to provide specific guidelines for working with the cell lines discussed but also to provide a conceptual framework and to suggest modifications and empiricisms that might be necessary in attempting to initiate studies using different cell lines.

2. USE OF ISOLATED MEMBRANE VESICLES OF ENTERIC BACTERIA FOR THE STUDY OF TRANSPORT

2.1. Background and Rationale

The mechanistic principles of bacterial transport have been defined by a number of biochemical, biophysical, and genetic approaches. Recently acquired knowledge of membrane architecture will provide further impetus for the inevitable merger of structural and functional insights concerning transport mechanisms. Despite continued progress, fundamental questions, such as the way the cell provides energy for nutrient uptake, have not been satisfactorily answered. Part of the difficulty stems from past reliance on intact cells, where membrane-associated transport processes are often obscured by metabolic contributions of the cytoplasm. Dissociation of these two components theoretically can be achieved by preparing isolated membrane vesicles. This approach came to fruition in 1966 when Kabāck and Stadtman successfully prepared membrane vesicles from *Escherichia coli* which could catalyze the active uptake of proline. Since that time, membrane vesicles derived from a variety of bacteria have been shown to accumulate such substrates as amino acids (Kaback and Milner, 1970; Lombardi and Kaback, 1972; Short *et al.*, 1972*a,b*; Konings and Freese, 1972), sugars (Barnes and Kaback, 1970; Kaback, 1972; Barnes, 1973), nucleic acid precursors (Hochstadt-Ozer and Stadtman, 1971*b*; Hochstadt-Ozer, 1972; Hochstadt-Ozer and Cashel, 1972; Komatsu and Tanaka, 1973; Pickard *et al.*, 1974; Hochstadt, 1974), and ions (Barnes, 1974; Bhattacharyya, 1970; Lombardi *et al.*, 1973). These metabolites are accumulated against a concentration gradient by metabolically dependent processes (Kaback and Hong, 1973; Kaback, 1972) or by group translocation (Kaback, 1968; Hochstadt-Ozer and Stadtman, 1971*b*).

The purpose of the following discussion is to highlight the remarkable insights into transport mechanisms which bacterial vesicles have provided. The vesicles are prepared by osmotic lysis of spheroplasts (or protoplasts, in the case of gram-positive cells) with the concomitant release of their cytoplasmic content. It is the loss of the intracellular enzymes and metabolite pools which makes the bacterial vesicles so valuable for transport studies. Prior to the use of bacterial vesicles, uptake of a metabolizable substrate by an intact bacterial cell would result in metabolic modifications leading to its subsequent incorporation into macromolecules or its loss through oxidation to carbon dioxide production. In order to avoid this, the investigator would have to resort to a nonmetabolizable substrate, obtain mutants with appropriate enzymatic blocks, or use inhibitors of macromolecular biosynthesis. None of this is necessary when vesicles are used for transport studies since the vesicles are essentially devoid of intracellular enzymes and intermediary metabolites. Some metabolic alterations of the substrate may occur, however, due to the presence of membrane-associated enzymes. Such vesicle-mediated alterations of substrate have aided in the understanding of an entirely new transport process, referred to as "group translocation." In contrast to classical active transport, the substrate is metabolically converted to a new derivative as it is translocated from one side of the membrane to the other. For example, a glucose molecule in the process of being transported is preferentially phosphorylated relative to any free glucose already in the vesicle (Kaback, 1968). Similar anisotropic processes account for the transformation of adenine (Hochstadt-Ozer and Stadtman, 1971b) and fatty acids (Frerman and Bennett, 1973) to AMP and fatty acyl CoA derivatives, respectively, as they are transported across certain bacterial membranes.

Actively transported substrates, in contrast to those which are group-translocated, remain metabolically unaltered; their accumulation against a concentration gradient requires energy which is presumably available in intact bacteria through the mediation of oxidative processes, hydrolysis of ATP, or an energized membrane state (Klein and Boyer, 1972; Berger, 1973). In order to identify the precise energy source, it is necessary to have a system which is responsive to exogenous energy supplies. Intact bacterial cells often do not fulfill this requirement because of the difficulty in adequately lowering their endogenous energy reserves (for an exception in a group translocation by "starved" cells, see Hochstadt-Ozer and Stadtman, 1971c). These energy reserves are essentially absent from isolated membranes, a property which makes them convenient for probing the sources of energy for active transport and the mechanisms whereby the energy is ultimately transferred to the transport carriers. This may not apply to group translocation systems because endogenous energy for these mechanisms is more readily available (e.g., 2-phosphoglycerate and P-ribose-PP) (see Kaback, 1968; Hochstadt-Ozer and Stadtman, 1971b).

The use of isolated membrane vesicles has shown that for much of classical active transport, substrate uptake depends on the passage of electrons down the respiratory chain (Kaback and Milner, 1970; Barnes and Kaback, 1970; Kerwar et al., 1972; Kaback, 1972; Konings and Freese, 1972; Short et al., 1972b). The oxidation of a variety of electron donors can be coupled to transport, but apparently there is a preferred substrate (depending on growth conditions) for each bacterium which is most efficient in stimulating uptake of metabolites (Kaback, 1972; Konings and Freese, 1972). While aerobic bacteria can generate energy for transport through several alternative pathways (Klein and Boyer, 1972; Berger, 1973), anaerobic bacteria were thought to be limited to substrate-level phosphorylations (Harold, 1972). This idea may no longer be tenable, since vesicles derived from E. coli cultured anaerobically contain transport systems that are coupled to electron flow but are not stimulated by ATP (Konings and Kaback, 1973).

The evidence for a close association between oxidative and active transport processes is substantial, but there is still controversy over the involvement of high-energy phosphate compounds. In vesicles, respiration-driven transport is not significantly diminished in the presence of arsenate and ATPase inhibitors (Kerwar *et al.*, 1972; Short *et al.*, 1972*b*; Kaback and Hong, 1973; Hirata *et al.*, 1971). Furthermore, transport properties are not substantially different in "wild-type" membranes or in vesicles prepared from mutants in which transport is uncoupled from oxidative phosphorylation (Prezioso *et al.*, 1973). The evidence seems to suggest that neither the production nor the utilization of high-energy phosphate compounds is linked to active transport. On the other hand, the membrane preparatory procedures may lead to loss of essential factors which are necessary to couple the ATPase to respiration and transport. This is not inconceivable since the respiratory chain itself is easily dissociated from transport by excessive washing of vesicles obtained from anaerobically grown *E. coli* (Konings and Kaback, 1973). Recently, the ATPase molecule has been implicated in the energy-coupling step. Vesicles prepared from some ATPase mutants will not transport substrates in the presence of an electron donor (Van Thienen and Postma, 1973; Rosen, 1973; Simoni and Shallenberger, 1972). It has been suggested (Rosen, 1973) that apart from its enzymatic activity the ATPase protein may act as an energy transducer between the respiratory chain and the transport carriers. The presence of enzymatically inactive ATPase proteins would still allow transport to occur, but the actual absence of these proteins, whether enzymatically active or not, could lead to an inability to concentrate substrates.

Several laboratories have suggested that an energized membrane state may supply the driving force for transport (Harold, 1972; Berger, 1973; Klein and Boyer, 1972; Van Thienen and Postma, 1973). This energy-conserving device may be synonymous with the proton gradient which is established during passage of electrons down the respiratory chain. According to the chemiosmotic hypothesis, oxidation of substrates leads to ejection of protons from cells or membrane vesicles with the production of a pH gradient (inside alkaline) and an electrical potential gradient (inside negative). The resultant protonmotive force causes active accumulation of metabolites. A common carrier is presumably available for both the protons and the specific substrate. The movement of protons "down" the concentration gradient results in a simultaneous transport of nutrients "up" a concentration gradient. The existence of such a mechanism in bacterial cells has been demonstrated by observing the correlation between galactoside uptake and changes in the membrane potential and pH gradient (Kashket and Wilson, 1973). In addition, Reeves (1971) has confirmed that an external proton gradient is established during oxidation of D-lactate by *E. coli* membrane vesicles. Recently, it was reported that an artificially imposed membrane potential, in the absence of electron transport, can by itself drive the active uptake of sugars and amino acids into vesicles (Hirata *et al.*, 1974). The protonmotive force may thus serve as an indirect link between respiration and transport. However, rubidium uptake in the presence of valinomycin seems to occur independently of a proton gradient but is coupled to electron transport (Lombardi *et al.*, 1973). As an alternative to the chemiosmotic hypothesis, it has been suggested (Kaback, 1972) that a segment of the electron transport system may be directly coupled to transport. This latter model includes the possibility that the electron carriers may simultaneously serve as transport carriers.

While the mechanism(s) of energy coupling remains controversial, it is apparent that the use of membrane vesicles has led to fresh insights and will continue to contribute to our understanding of active transport processes. This is only one example of how the use of isolated membrane vesicles contributed to widened conceptual understanding—in this case, to energy coupling in "classical active transport." Another instance in which use of membrane vesicles helped elucidate a transport mechanism was in the adenine group translocation system. [Although the group translocation of glucose was confirmed by use of vesicles (Kaback, 1968), the system had already been well characterized by a variety of biochemical and genetic techniques (see Roseman, 1969).]

Adenine uptake by *E. coli* is mediated by membrane-localized adenine phosphoribosyltransferase, which results in the accumulation of intravesicular AMP; this process is dependent on phosphoribosylpyrophosphate (Hochstadt-Ozer and Stadtman, 1971*b*). The enzyme is largely extractable from the pericytoplasmic space by osmotic shock (Hochstadt-Ozer and Stadtman, 1971*c*). The kinetic and regulatory parameters of this enzyme as a soluble protein purified to homogeneity (Hochstadt-Ozer and Stadtman, 1971*a*) when localized on isolated membrane organelles (Hochstadt-Ozer and Stadtman, 1971*b*) and when in the intact cell (Hochstadt-Ozer and Stadtman, 1971*c*) are virtually identical.

The uptake of adenosine was found to occur in two steps: (1) initial cleavage of adenosine to adenine and (2) uptake of the adenine by the same mechanism as for free adenine (Hochstadt-Ozer, 1972). In addition, it was learned that 5'-nucleotides are capable of regulating the transport process and that the overall regulation of uptake of purines (Hochstadt-Ozer and Stadtman, 1971*b*) and pyrimidines (Hochstadt-Ozer and Cashel, 1972) by ppGpp (Hochstadt-Ozer and Cashel, 1972) seems to play an important role in the phenomenon of amino acid control of RNA metabolism. Furthermore, substrate turnover rates in cells made dependent on exogenous purines indicate that most, if not all, of the cellular adenine phosphoribosyltransferase must be at or near the membrane *in situ* and participating in uptake in order to account for the observed transport and growth rates (Hochstadt-Ozer and Stadtman, 1971c).

Finally, membrane vesicles have also been effectively used to demonstrate a third type of mediated transport, facilitated diffusion (Kaback and Stadtman, 1968). A facilitated diffusion transport system depends on a specific carrier, shows saturation kinetics, and is inhibited by inactivation of the protein carrier. It neither requires energy nor results in concentration of substrate against a gradient. A facilitated diffusion system is thus a carrier-associated means of rapidly equilibrating substrate concentrations on both sides of a membrane vesicle.

2.2. Preparative Techniques

The production of membrane vesicles to study such transport mechanisms and other membrane activities depends on initial conversion of the bacteria to osmotically sensitive spheres by disruption or removal of the outer layers of the cell envelope. The precise stratagem employed depends on whether the organism is gram positive or gram negative, since the bacterial envelope differs in these two classes. The surface structures of a gram-positive organism consist of a peptidoglycan cell wall surrounding a cytoplasmic membrane. By use of various enzymes (e.g., lysozyme for *Micrococcus* and *Bacillus* species or lysostaphin for *Staphylococcus* species) it is possible to completely remove the cell wall, thereby producing osmotically sensitive spheres referred to as "protoplasts." The cell envelope of gram-negative, enteric bacteria differs from that of the gram-positive organisms by an additional membrane external to the peptidoglycan layer. This outer membrane is comprised of protein, phospholipid, and most of the lipopolysaccharide of the cell envelope (Osborn *et al.*, 1972). A chelating agent is used to disrupt the outer membrane in order to expose the underlying peptidoglycan layer to the action of lysozyme. The resultant osmotically sensitive form is termed a "spheroplast" as a consequence of the incomplete removal of the cell envelope. Exposure of protoplasts or spheroplasts to hypotonic conditions leads to lysis and the conversion of the cytoplasmic membranes into closed vesicles essentially devoid of cytoplasmic enzymes, metabolic intermediates, and other cytoplasmic constituents.

Much of the work in our laboratory has concentrated on enteric bacteria and therefore we shall emphasize the methods used for preparation of membrane vesicles from these organisms. The reader is referred to Kaback's (1971) review on methodology for supplemental information. Membrane vesicles have also been isolated from *Mycobacterium phlei* (Hinds and Brodie, 1974), a marine pseudomonad (Sprott and MacLeod, 1972, 1974), *Azotobacter vinelandii* (Barnes, 1972), *Bacillus licheniformis* (MacLeod *et al.*, 1973), *Bacillus subtilis* (Konings and Freese, 1972; Konings *et al.*, 1973; Hampton and Freese, 1974), and *Staphylococcus aureus* (Konings *et al.*, 1971). It should be noted that certain procedural modifications are necessary in order to prepare vesicles from anaerobically grown *E. coli* (Konings and Kaback, 1973).

2.2.1. For Many *E. coli* Strains

E. coli membrane vesicles are prepared essentially according to the method of Kaback (1971). The procedure involves washing the cells, converting them to osmotically sensitive spheres with lysozyme-EDTA, lysing these spheroplasts in hypotonic buffer, and washing the resultant membrane vesicles. Although it has been described elsewhere (Kaback, 1971), further details about the procedure, particularly as it is utilized in this laboratory, may be of interest.

1. Cells are collected by centrifugation and the drained cell pellets may be left in the refrigerator overnight. In our experience, storing the cells frozen prior to membrane preparation leads to loss of activity of nucleic acid precursor transport systems, although other transport systems may remain active. Cells are washed twice with approximately 30 vol or more of 10 mM tris-HCl (pH 8.0). The tubes for the centrifugation are preweighed in order to facilitate the next step.

2. The cells are resuspended at a dilution of 1 g of cells per 80 ml of "spheroplasting" medium; the "spheroplasting" medium is 30 mM tris-HCl (pH 8.0), 20% sucrose, 10 mM potassium EDTA (pH 7.0), and 0.5 mg/ml lysozyme. Cells are stirred in the "spheroplasting" medium for 30 min at ambient temperature. This process should be monitored by phase contrast microscopy. Some lysozyme-sensitive strains "round up" and assume the spheroplast shape. However, only a small proportion of commonly studied strains of enteric bacteria will be morphologically altered under these conditions. Strains that do not change shape during this treatment will frequently give a high yield of membrane vesicles during the subsequent lysis step. It is usually convenient to perform this second step in a very large beaker or carboy set on a magnetic stirrer. Most strains which do not "round up or" swell during lysozyme treatment may, if necessary, be made more lysozyme sensitive by the following modifications:

a. Raising the pH to 9.0–9.4 (retention of transport activity will determine the advisability of this treatment).

b. Raising the temperature from 20°C to 30–37°C.

c. Increasing the incubation time from 30 min to 1–3 h.

d. Using the alternative method (see Section 2.2.2), which is a modification of that of Osborn *et al.* (1972).

3.   The spheroplasts are collected by centrifugation and immediately resuspended in the smallest convenient volume of 0.1 M potassium phosphate buffer (pH 6.6) containing 20% sucrose, 20 mM $MgSO_4$, and 3–5 mg/ml DNase I (Worthington); 3–5 mg RNase is optional. This step is performed in a 100–200 ml teflon and glass homogenizer (mechanically assisted; vigorous homogenization is helpful in preparing a uniform suspension). If possible, the final volume should be limited to 50–100 ml. Because high viscosity and cell–DNA aggregates may make homogenization difficult, the pellet should be resuspended immediately.

4.   The resuspended spheroplasts are warmed to 37°C and slowly added, with stirring, to 300–500 vol of 50 mM potassium phosphate buffer (pH 6.6); the latter solution is referred to as "lysis medium." This step is performed in a large beaker or carboy with the lysis medium preincubated at 37°C. The large dilution volume used to lyse the spheroplasts results in the formation of membrane vesicles containing as little as 0.2% activity of cytoplasmic enzymes (e.g., glutamine synthetase; see Hochstadt-Ozer and Stadtman, 1971c). However, since certain transport systems and their components have been found to be less tightly bound to the membrane vesicles than others, it may be necessary to use a smaller dilution volume when lysing the spheroplasts. Such a modification results in somewhat greater cytoplasmic contamination, but also greater retention of transport activity. For some transport systems, it has been useful to prepare membrane vesicles by shocking the spheroplasts into only 10–20 vol of the buffer rather than into 200–500 vol. After the addition of the spheroplasts to the lysis medium, incubation proceeds at 37°C with rapid stirring, for approximately 15 min.

5.   An appropriate amount of potassium EDTA (usually 1 M solution at pH 7.0) is added to bring the EDTA concentration in the lysis medium to 10 mM. Incubation with stirring at 37°C is continued for 15 min or more.

6.   1 M $MgSO_4$ is added to the lysis medium to bring the final concentration of $MgSO_4$ to 15 mM (disregarding the amount bound to EDTA). Incubation is continued at 37°C with stirring for 15 min or more. The lysis is monitored by phase contrast microscopy. More than 90% of the rods

and spheroplasts (spheres with dark or dense appearance) should be converted to vesicles (swollen spheres with transparent appearance). If this is not observed, modifications delineated in step (2) may be necessary. The degree to which digestion of released DNA occurs is indicated by the reduction in viscosity of the solution on addition of $MgSO_4$. If much undigested DNA is observable, the incubation times may be increased.

7. Membrane vesicles, unlysed spheroplasts, and intact cells are collected by centrifugation at 9000g for 1 h in 500-ml bottles using a Sorvall GS3 rotor (a continuous-flow rotor may also be used for very large volumes). (This and all subsequent g-force values refer to the force exerted at the center of the centrifuge tube.) Centrifuge brakes should not be used and care should be exercised in decanting supernatant fluids as the pellets are often very loosely packed. High-speed, large-capacity swinging-bucket rotors are useful for this step. Differential centrifugation, as described below, will lead to recovery of 50–70% of the vesicles. Repeated cycles of the centrifugation procedure will increase the yield. The particulate material is resuspended in a small volume of 0.1 M potassium phosphate buffer (pH 6.6) containing 10 mM potassium EDTA (unless EDTA addition at this step is found to be deleterious to final transport activity). Resuspension is aided by a motor-driven teflon and glass homogenizer or by vigorous pipetting with a pasteur-type pipette. The uniform suspension is then diluted with 20–100 vol of the phosphate–EDTA buffer and centrifugation is performed at 800g for 20–30 min. Intact cells should pellet and membrane vesicles remain in the supernatant fluid. Each supernatant fraction is examined under phase contrast microscopy and those with 1% or fewer rod forms are pooled as the vesicle fraction. Since several cycles of the low-speed centrifugation step will increase the final membrane yield, this aspect of the procedure is repeated on the resuspended pellet until the supernatant fluid is relatively clear compared to the original membrane suspension. If the supernatants from this differential centrifugation procedure do not contain a cell-free population of membrane vesicles, the sucrose barrier technique (Kaback, 1971) may be employed. The pooled supernatant fluids are centrifuged at 45,000g for 20 min in order to collect the membrane vesicles. The pellet is washed once in 0.1 M phosphate buffer (pH 6.6) using a teflon and glass homogenizer and resuspended in 0.1 M potassium phosphate buffer (pH 6.6) at a concentration of approximately 10 mg membrane vesicle protein per milliliter. This concentration corresponds to an approximate optical density of 1.50 at 660 nm.

8. The vesicles are frozen in liquid nitrogen in aliquots of 1–2 ml and can then be stored in either liquid nitrogen or in a freezer at $-79°C$.

Membrane vesicles appear to be quite stable if frozen and stored under these conditions and show little loss of activity over several months or more. In general, storage at pH 6.5–7.5 seems to be innocuous, although storage at a pH outside this range may alter transport activity. Membrane vesicles can be collected by centrifugation at $\geq 20,000g$ and resuspended in alternative buffers, if necessary. To retain transport activity, the membranes must be thawed at 30–48°C depending on the transport system, with vigorous shaking, and placed in ice immediately. Membrane vesicles may be stored for several days or longer at 0–4°C, but must be monitored by phase contrast microscopy before use to insure freedom from bacterial contamination. Membranes that are to be used for transport assays should not be refrozen.

2.2.2. For Other *E. coli* and *Salmonella* Strains

While the procedure devised by Kaback (1971) outlined in Section 2.2.1 works rather well, it is not universally applicable. *Salmonella typhimurium* LT2 is relatively resistant to the EDTA-lysozyme treatment. As an alternative, it is our experience that a modification of a procedure developed by Osborn *et al.* (1972) will yield satisfactory quantities of membrane vesicles from this organism. These membrane vesicles are competent for nucleic acid precursor uptake and may be useful for the study of other transport systems.

S. typhimurium cells are grown in 2 liters of a defined medium on a rotary shaker at 37°C until the late log or early stationary growth phase. Harvesting is accomplished by centrifugation for 15 min at 13,000g. The centrifuge bottles are inverted to drain as much of the residual growth medium as possible and then the cell pellets are transferred to a teflon and glass homogenizer using a curved spatula. A minimal amount of cold 10 mM tris-HCl buffer (pH 7.8) containing 0.75 M sucrose is added and the cells are briefly homogenized to achieve a uniform suspension. Subsequent resuspensions are achieved by pipetting rather than homogenization. The resultant suspension is added to 320 ml of the sucrose–tris buffer at 4°C, which is gently stirred with a magnetic stirrer. Thirty-two milligrams of lysozyme (thrice recrystallized) is added, and, after a 2-min incubation at 4°C, 640 ml of cold 1.5 mM NaEDTA (pH 7.5) is added at the rate of approximately 100 ml/min. This dilution step usually results in excellent spheroplast formation, which is monitored by phase microscopy. For convenience, the spheroplast suspension is divided into two portions. One-half of the suspension is slowly added to 3 liters of distilled water (at ambient temperature), which is rapidly mixed by means of a magnetic stirring bar. Then 6 mg $MgCl_2$, 40 mg DNase, and 20 mg RNase are added and the suspension is incubated at ambient temperature for 20 min. The suspension is then centrifuged at 13,000g for 15 min. The second half of the original suspension is treated in the same manner. The combined pellets obtained from the distilled water lysis step are resuspended in 0.1 M potassium phosphate buffer (pH 7.0) and recentrifuged at 17,000g for 15 min. The pellets appear tannish in color with two or three distinct zones distinguishable by their coloration. The upper layer is comprised of membrane vesicles, while the lower layers contain a mixture of vesicles, cells, and spheroplasts. A preliminary separation of vesicles from unlysed contaminants is helpful. To accomplish this step, a minimal amount of the phosphate buffer is added to the tube and the upper zone is resuspended with gentle vortexing. The resuspended upper layer is placed into a separate tube and the rest of the pellet is then redispersed in phosphate buffer. These two tubes are centrifuged at 3000g for 15 min in order to remove whole cells and spheroplasts. Some loss of membrane vesicles into the resultant pellet is accepted in order to assure a purified vesicle population essentially free of cells and spheroplasts. The supernatant fluid is carefully decanted to avoid admixture with the pellet and is then centrifuged at 27,000g for 15 min. The membranes are washed in 0.1 M potassium phosphate buffer (pH 7.0), frozen in liquid nitrogen, and stored at $-79°C$. This method has the advantage of considerable reduction in the fluid volume of the lysis suspension as compared to the Kaback procedure (Kaback, 1971), and therefore less time is usually required to collect the vesicles by centrifugation. Attempts to replace distilled water as the lysing medium with 1 mM $MgSO_4$ or 50 mM phosphate buffer were not successful. This procedure, as an alternative to the Kaback procedure, was found essential for consistent preparation of membranes in high yield from certain strains of *Salmonella*.

While vesicles are valuable for transport studies, there are several problems which must be considered. Both pericytoplasmic and membrane components are released during the preparation of membranes. This may result in the loss of certain proteins and other components essential for transport function. Thus conclusions based on vesicle studies may not always be applicable to physiologically intact cells.

2.3. Membrane Sidedness

The orientation of membrane vesicles has always been assumed to be the same as in the intact parent cell since the vesicles take up various substrates (Kaback, 1972) and extrude protons (Reeves, 1971). Studies by Futai (1974*a,b*) and Weiner (1974) have led to a reconsideration of this concept. Their results demonstrate that several dehydrogenases and adenosine triphosphatase are internally located in spheroplasts and whole cells. These enzymes become partially accessible to the external medium in vesicles prepared by osmotic lysis, possibly because of the production of a mixed population of vesicles having both inside-in and inside-out orientations. However, freeze-fracture studies have shown that the particle distribution on the exposed faces of vesicles is the same as found on the cytoplasmic membrane of intact cells, indicating an outside-out orientation for the vesicles (Konings *et al.*, 1973; Altendorf and Staehelin, 1974). An alternative explanation for the altered enzyme localization in vesicles prepared by osmotic lysis might be that a small fraction of the enzymes move to the outside during vesicle formation (Futai, 1974*a,b*). Inverted or inadequately sealed vesicles, while comprising only a small fraction of a given membrane population, may play a role in determining the relative efficiency of various energy sources for driving transport. These vesicles may oxidize electron donors at a faster rate than closed right-side-out vesicles, but only the latter can accumulate substrates within their intravesicular space. Hampton and Freese (1974) have pointed out that this could lead to the misinterpretation that some energy sources are rather inefficient. After appropriate corrections are made to compensate for the presence of inverted or leaky vesicles, it now appears that various electron donors are probably equivalent in their ability to stimulate transport. Attempts have been made to reconstitute active transport systems by trying to bind purified dehydrogenases to membranes derived from cells deficient in these enzymes (Futai, 1974*a*; Short *et al.*, 1974; Reeves *et al.*, 1973). While the experiments have proven successful, there are differences in the response of the reconstituted vesicles to various inhibitors compared to membranes from wild-type cells.

The considerations set forth in this section should not detract from the vast amount of knowledge which has been gained from vesicle experimentation. It should be readily apparent that a fuller appreciation of transport mechanisms will continue to depend on the use of vesicles in transport studies.

2.4. Transport Assay

Membranes prepared as described above are assayed for transport activity as follows.

Fifty microliters of distilled water is added to 12- by 75-mm disposable glass test tubes; Hamilton syringes in Hamilton repeating dispensers are used to disperse reagents into the reaction tubes. Effectors that putatively or potentially act from within the vesicle (at the inner membrane surface) are then added to the water at final concentrations of 20 mM or less (e.g., phosphoenolypyruvate for assay of glucoside transport; see Kaback, 1968). The membranes, suspended in 0.1 M potassium phosphate buffer at an appropriate pH, are then added to the tubes at a dilution of 1:2. This provides for the possibility of osmotically shocking the effector or cofactor into the vesicles by a transient osmotic lysis. After dilution of the membranes, other effectors, cofactors, and salts are added to the tubes.

The reaction mixture (0.1–0.125 ml total volume) is preincubated at an optimal temperature for 5–15 min and the reaction is initiated by the addition of the radioactively labeled substrate to be transported. Transport kinetics can be measured by varying the time of initiation of the reaction so that all samples in an assay are terminated at the same time. However, for short incubations times (2 min or less) each sample should be assayed separately. The optimal temperature for transport will depend on the substrate used, the growth conditions of the culture from which the membranes were derived, and even on the time of incubation. That is, initial transport rates (5 min or less) may be optimal at temperatures above 40°C while steady-state uptake (20 min or longer) may be optimal at 30–40°C (see Hochstadt, 1974; Hochstadt-Ozer, 1972; Kaback, 1968). At the end of the incubation period, the reaction mixture is rapidly diluted with 2 ml of 0.5 M NaCl (kept at the assay temperature); the membranes are then collected on 0.45-μm nitrocellulose filters by suction. The background control for each assay is a sample to which the radioactive substrate is added after dilution with 0.5 M NaCl and then immediately filtered. The filters are washed twice with 2-ml aliquots of salt solution. Approximately 30 s is required to terminate the reaction and remove the filters from the vacuum apparatus. The filters are then placed on an absorbent surface (e.g., paper towels), from which they are transferred to planchets (or scintillation vials). They are secured on the planchet with a metal ring, dried (e.g., 2 min at 120°C), and monitored for radioactivity. Gas flow counting, when available, has the advantage of economy in both time and material since the same sample can be used for elution and chromatographic analysis of vesicle contents. Analysis of vesicle contents for uptake products of nucleic acid precursors is described in Section 3.3.2i.

3. TRANSPORT STUDIES USING MEMBRANE VESICLES ISOLATED FROM CULTURED MAMMALIAN CELLS

3.1. Rationale

It has been proposed by Holley (1972) that control of normal cell growth and neoplasia may be determined by alterations in the plasma membrane. Such alterations may directly or indirectly involve transport function or its regulation. One of the interests of this laboratory is to determine whether growth in normal and neoplastically transformed mammalian cells in culture is regulated by differences in specific membrane transport systems—in particular, uptake of purine and pyrimidine bases and their nucleosides. Certain mutant cell lines (e.g., thioguanine resistant and ouabain resistant) are also employed to better analyze specific functions.

Until recently, most studies of mammalian cell transport have dealt with intact cells, in which it is difficult to distinguish, using metabolically active substrates, between membrane transport events and subsequent events occurring during intracellular, intermediary metabolism. To better understand the mechanism, regulation, and overall function of a transport system, it is most advantageous to use isolated membrane vesicles.

3.2. Background

The assay of transport activity in isolated membrane vesicles derived from mammalian cells represents the final step in a series of critical and largely empirical procedures. The preparation of isolated membranes possessing transport activity can be divided into five sequential steps: selection of "tissue," choice of membrane and organelle "markers," cell disruption, cell fractionation, and transport assay. Although a general discussion of these factors will follow, emphasis will be placed on the use of mammalian cells grown in culture.

Problems associated with the choice of tissue as a cell source have been reviewed by DePierre and Karnovsky (1973). Tissue contains several cell types which can rarely be separated successfully. A homogeneous cell population is desirable in order to obtain plasma membranes from a single cell type. This can be achieved with tissue culture cell lines, but plasma membranes isolated from cloned, cultured cells may still demonstrate heterogeneity due to membrane mosaicism (Wallach, 1967) or cell-cycle changes (Noonan *et al.*, 1973; Graham *et al.*, 1973).

The next step involves the selection of suitable membrane markers in order to follow plasma membrane isolation. The selection of suitable markers involves some potential pitfalls, which have been stressed in several reviews (Wallach and Lin, 1973; Lauter *et al.*, 1972; DePierre and Karnovsky, 1973). It must be emphasized that it is not sufficient to utilize a marker only for the specific membrane of interest. Since the final purity of the membrane preparation is important, markers for all subcellular organelles are necessary in order to ascertain levels of contamination. Selection of markers for the plasma membrane generally involves enzymes "exclusively" localized on the plasma membrane. Examples of such enzymes include 5'-nucleotidase (Allan and Crumpton, 1970; Quinlan and Hochstadt, 1974), Na^+/K^+-activated Mg^{2+}-dependent adenosine triphosphatase (Gahmberg and Simons, 1970; Graham, 1972; Wallach and Kamat, 1966b), and adenyl cyclase (McKeel and Jarrett, 1970). The selection of marker enzymes is largely empirical and should be validated independently for each system (Neville, 1975). Thus, while 5'-nucleotidase has generally been assumed to be a good plasma membrane marker, there are cell types in which it cannot be used, e.g., rat adipocytes (McKeel and Jarrett, 1970) and Ehrlich ascites tumor cells (Wallach and Kamat, 1966b). One must remember that cytochemistry was often the original basis for designating a particular enzyme as a plasma membrane marker and this method of localizing enzyme activity is known to generate numerous artifacts (DePierre and Karnovsky, 1973). Alternatives to the use of marker enzymes are varied, including chemical markers (fucose), plasma membrane iodination (Wallach and Lin, 1973), and serological techniques (Ozer and Wallach, 1967).

As previously stated, it is also necessary to select markers for the various subcellular organelles in order to determine the degree of membrane purity. Wallach and Lin (1973) and Neville (1975) have presented and discussed a list of enzymes that may be used as markers for mitochondria, endoplasmic reticulum, lysosomes, and Golgi membranes. However, the validity of using some marker enzymes will depend on whether a particular organelle remains intact during the isolation procedures. If the marker is an intraorganelle enzyme, then the problem of substrate permeability must be considered. If the marker enzyme is located, for example, on the outer mitochondrial membrane it is useful only if cell disruption and subsequent fractionation do not lead to the loss of that outer membrane. It is best to use marker enzymes for both inner and outer mitochondrial membranes, for if mitochondria are disrupted the outer membrane may contaminate the plasma membrane fraction.

The selection of a series of membrane and organelle markers to be utilized during cell fractionation and membrane isolation is followed by the process of cell disruption. One of the first methods reported (Neville, 1960) involved Dounce homogenization of rat liver tissue in hypotonic medium in the absence of divalent cations. The yield of plasma membranes was less than 1%, while more than 90% of the nuclei lysed. Any cell disruption procedure involving a form of osmotic lysis increases the probability that subcellular organelles will also lyse. The low yield of plasma membranes obtained with the above procedure was improved on by the addition of calcium to the disruption medium (Ray, 1970). McKeel and Jarrett (1970), using rat adipocytes, included sucrose in their homogenizing medium in order to prevent excessive lysis of subcellular organelles. Other approaches designed to improve the yield of plasma membranes have been reported. Using mouse fibroblasts (L cells), Warren and his associates (Warren *et al.*, 1966) added stabilizing agents such as zinc salts and fluorescein mercuric acetate to the Neville disruption medium. The use of such stabilizers causes the formation of large sheets of plasma membranes and "ghosts" and thus facilitates the separation of the plasma membranes from smaller cell "debris." However, heavy metal stabilizers could have adverse effects on enzymes and on membrane functions such as transport. One promising technique which would simultaneously increase membrane yield and purity is the use of affinity columns directed against specific surface components. Soderman *et al.* (1973) have prepared agarose columns to which insulin is covalently bound. Since the plasma membrane contains almost all of the insulin receptors, the plasma membrane is selectively retained when disrupted cellular suspensions are passed through these columns. It is potentially possible to use antibody directed against certain H-2 antigens localized exclusively on the plasma membrane (Ozer and Wallach, 1967) of a wide variety of cell types in order to have columns of more generalized specificity.

In order to study transport at the level of the isolated membrane, one needs closed, selectively permeable membrane vesicles; however, many of the methods of cell disruption discussed above result in the formation of membrane sheets or ghosts containing large holes. Modifications of the method of McKeel and Jarrett (1970) have been successfully applied to the production of closed plasma membrane vesicles for transport studies (Carter and Martin, 1969; Carter *et al.*, 1972; Li and Hochstadt, 1975). Another cell disruption method providing closed vesicles is the nitrogen-cavitation procedure (Wallach and Kamat, 1966*b*), in which cells are equilibrated in a nitrogen atmosphere, usually at 600–800 lb/inch². When the cells are released from the high pressure, gas bubbles form at the plasma membrane and cell rupture results, possibly due to liquid shear (Wallach and Lin, 1973). The advantages of this method are as follows:

a. A variety of cell types are susceptible to quantitative disruption.
b. Cell disruption is performed in an inert atmosphere, using an isotonic buffer, and under conditions which allow the temperature to be controlled.
c. No local heating occurs as would be produced during sonic disruption.
d. Subcellular organelles remain intact (Steck and Wallach, 1970; Gahmberg and Simons, 1970).
e. The isolated plasma membrane vesicles are closed and selectively permeable (Wallach and Kamat, 1966a; Wallach et al., 1966), which makes them suitable for transport studies.

Following cell disruption, fractionation procedures are performed to separate plasma membranes from subcellular organelles. The Neville procedure (Neville, 1960), which depends on differential centrifugation and sucrose velocity gradients to isolate the plasma membrane, has been employed by several investigators (Ray, 1970; Warren et al., 1966). A major advance in cell fractionation procedures was developed by Soderman et al. (1973), who used insulin–agarose columns to isolate the plasma membrane.

In addition, Brunette and Till (1971) introduced an aqueous, two-phase polymer system of dextran and polyethylene glycol to purify L-cell plasma membranes. Their method, however, utilizes zinc salt as a stabilizing agent and yields membrane sheets and ghosts. Advantages include the high yield of plasma membranes (more than 50% based on marker enzyme analysis) and the relatively short preparation time (2–3 h) required. The two-phase polymer system was also utilized by Lesko et al. (1973) to isolate liver cell plasma membranes and by Gruenstein et al. (1974) to isolate plasma membranes from HeLa cells.

The nitrogen-cavitation method of cell fractionation developed by Wallach and Kamat (1966b) yields a mixed vesicle population of plasma membranes and endoplasmic reticulum (collectively referred to as "microsomes" because of their sedimentation properties). The two membrane populations can be separated from each other using ficoll or dextran gradients. The theory underlying the separation of plasma membranes and endoplasmic reticulum, which is based on differences in membrane surface charge, has been presented elsewhere (Steck and Wallach, 1970; Wallach and Lin, 1973). Resolution is favored when gradients are utilized that possess low osmotic (dextran or ficoll) and ionic activities in the presence of a low concentration of divalent cations; the latter apparently lowers the fixed charge on the membrane. Although one obtains plasma membrane vesicles of high purity that are competent for transport, conditions are used (low ionic strength washes) that may result in loss of certain membrane-bound molecules.

Before proceeding to a discussion of transport studies using membrane vesicles derived from eukaryotic cells, it must be stressed that it is essential to maintain total bookkeeping of marker enzyme activities following cell fractionation (Neville, 1975). It is not sufficient to isolate a plasma membrane fraction enriched 30- to 40-fold in marker enzyme specific activity. If the isolated plasma membrane accounts for only 5% of the total activity in the homogenate, where is the other 95% localized? If the bulk of the marker enzyme activity resides in another particulate fraction, perhaps the fractionation procedure can be modified in order to obtain a greater yield of plasma membrane. However, if most of the enzyme activity cannot be accounted for at all, then one must consider that that particular marker is inadequate, despite a high enrichment value in the putative plasma membrane fraction.

In general, studies of mammalian cell transport systems have used intact cells. Transport substrates examined include glucose and its analogues (Romano and Colby, 1973; Venuta and Rubin, 1973), nucleosides (Taube and Berlin, 1972; Plagemann and Erbe, 1972; Cunningham and Pardee, 1969), calcium ions (Perdue, 1971), and amino acids (Wheeler and Christensen, 1967). Uptake studies involving isolated plasma membrane vesicles have only recently been performed (Quinlan and Hochstadt, 1974, 1975*a, b*; Carter and Martin, 1969; Carter *et al.*, 1972; Li and Hochstadt, 1975), subsequent to the availability of selectively permeable plasma membrane vesicles (Wallach and Kamat, 1966*b*; McKeel and Jarrett, 1970).

The availability of membrane vesicles from normal, transformed, and mutant cell lines should prove of considerable value in clarifying the role of the membrane in various cellular processes. As an example, Cunningham and Pardee (1969) found that contact-inhibited Swiss 3T3 cells transported uridine at a significantly lower rate than did actively growing cells. Since intact cells were used, it was difficult to determine whether the change in transport rate was due to a membrane alteration(s) or to a change in an intracellular enzyme, such as uridine kinase. That the decreased uridine transport rate exhibited by contact-inhibited 3T3 cells was indeed due to a membrane-localized change(s) has been demonstrated by the use of membrane vesicles derived from these cells (Quinlan and Hochstadt, 1974). Benke *et al.* (1973) demonstrated that isolated, intact fibroblasts from Lesch-Nyhan patients (inactive hypoxanthine phosphoribosyltransferase) were unable to transport hypoxanthine or guanine. It has since been demonstrated (Quinlan *et al.*, in prep.), using membrane vesicles isolated from wild-type 3T3 cells, that hypoxanthine uptake does occur and is stimulated 10-fold by phosphoribosylpyrophosphate but that this stimulatory effect is not observed in membrane vesicles derived from thioguanine-resistant cells (inactive hypoxanthine phosphoribosyltransferase).

3.3. Nitrogen-Cavitation Technique for Isolation of Membrane Vesicles

A mixed vesicle population containing both plasma membrane and endoplasmic reticulum can be isolated from monolayer cultures of Balb/c 3T3 cells, SV40-transformed Balb/c 3T3 cells, baby hamster kidney (BHK) fibroblasts, and polyoma-virus-transformed BHK cells. The method of cell disruption and membrane isolation is a modification of the Wallach and Kamat (1966*b*) nitrogen-cavitation procedure. Briefly, "broken" cells are subjected to a series of differential centrifugation steps which yield a mixed vesicle population containing plasma membranes and endoplasmic reticulum.

Further fractionation to separate the plasma membrane and endoplasmic reticulum vesicles from each other can be accomplished with dextran gradients. The final yield of plasma membrane vesicles from 3T3 cells is increased significantly by releasing "trapped" plasma membrane vesicles from the mitochondrial fraction and pooling them with the original mixed vesicle population. The isolated mixed vesicle population is capable of transporting nucleosides by a mediated process and there is a significant increase in the specific activity of transport on isolation of the plasma membrane (Quinlan and Hochstadt, 1974, 1975*a*, *b*; Dowd and Hochstadt, manuscript in preparation).

3.3.1. Reagents Used

PBS: phosphate-buffered saline (50 mM potassium phosphate, pH 7.5, 0.15 M NaCl).

TSM: 5 mM tris-HCl (pH 7.5), 0.25 M sucrose, 0.2 mM $MgCl_2$ or $MgSO_4$.

TS: 5 mM tris-HCl (pH 7.5), 0.25 M sucrose.

TS-2: 1 mM tris-HCl (pH 7.5), 0.25 M sucrose.

Dextran I: dextran-110,* 10 mM tris-HCl (pH 8.0), 0.1 M sucrose, 1 mM $MgCl_2$.

Dextran II: dextran-110,* 1 mM tris-HCl (pH 8.0), 1 mM $MgCl_2$.

Dextran III: dextran-110,* 1 mM tris-HCl (pH 8.0), 0.2 mM $MgSO_4$.

3.3.2. Procedures

a. Cells and Growth Medium. Balb/c 3T3 cells and SV40-transformed Balb/c 3T3 (hereafter referred to as SV-3T3) are grown in 720-cm² roller bottles at 37°C. The cells are fed every 3 days with Dulbecco's modified Eagle's medium containing 1 mM glucose, 2 mM glutamine, and 10% fetal calf serum (final concentration). The initial pH of the growth medium is 7.5–7.6 and at harvesting never less than 7.3–7.4. Stock cultures can be maintained in 75-cm² flasks. The BHK-21 cells or polyoma-transformed BHK-21 cells (BHK-Py) are grown as above or in BHK-21 medium (GIBCO) containing 0.2% tryptose phosphate broth, supplemented with 10% fetal calf serum.

b. Harvesting of Cells. All subsequent centrifugation and wash steps are performed at 2–4°C. The 3T3 and SV-3T3 cells, while still attached to their substratum, are washed once with PBS (with or without $CaCl_2$), while BHK cells are washed twice with PBS alone. Harvesting is accomplished by scraping with a rubber policeman. The cells are collected as a pellet by centrifuging at 500g for 15 min and are resuspended in TSM and recentrifuged as above. After resuspension of the resultant pellet in TSM buffer, a centrifugation step (1000g for 10 min) is carried out using a 50-ml calibrated centrifuge tube. The packed cell volume is recorded (e.g., 15 roller bottles of SV-3T3 cell culture yields 5–6×10^5 cells/cm², for a total of 8–10 ml packed cell volume) and the cells are resuspended in 10 vol of TSM. At this stage, the cells should be well dispersed. This can be accomplished by passing them through a small-bore pipette three or four times or by gently homogenizing the cells with a loose-fitting Dounce homogenizer. The vigor and length of the resuspending procedure are adjusted to reduce cell clumps to single cells without loss of cell viability. These parameters should be monitored by phase contrast microscopy and dye exclusion; a cell count per unit volume is made at this time in a cell counting chamber slide.

c. Nitrogen-Cavitation Treatment. The nitrogen-cavitation procedure described by Wallach and Kamat (1966b) is used with the following modifi-

---

* The dextran-110 concentrations are specified in the text.

cations. The nitrogen pressure is lowered to 650–700 lb/inch^2 and the equilibration time is shortened to 10–15 min for 3T3 and SV-3T3 cells. For BHK cells, the optimal condition for cell breakage in subconfluent cultures is 600 lb/inch^2 for 5–8 min, and for confluent cultures it is 825 lb/inch^2 for 20 min. These modifications help prevent nuclear rupture, which can be further minimized by changing the growth medium 24 h or less prior to harvesting. Initially, only a few drops are expelled from the "bomb," and after waiting several minutes this released suspension is examined by phase contrast microscopy to ascertain that complete cell breakage has occurred and that there are as many intact nuclei as there were cells in the original suspension.

Although this is the appropriate procedure for most cell lines since rupture can be best observed only when some minutes have elapsed after release from the bomb, it is not suitable for all. BHK cells, for example, seem to rupture immediately, and waiting to observe the initial sample will seriously jeopardize nuclear integrity of the cells released later since time of incubation appears to be critical with these cells.

By performing a cell count prior to the cavitation procedure, and a nuclear count after, an estimate of nuclear rupture can be obtained. Excessive nuclear "breakage" can be lessened by use of a lower nitrogen pressure and/or a shorter equilibration time. Thus nitrogen pressure and equilibration may have to be adjusted relative to each growth condition and cell type in order to optimize cell breakage and minimize nuclear breakage. Once the appropriate conditions are ascertained, the remainder of the cells are released from the "bomb" and potassium EDTA (pH 6.9) is added to the suspension at a final concentration of 1 mM to prevent particle aggregation.

d. Differential Centrifugation. The total cell homogenate is then centrifuged at 500–800g for 10–15 min in order to separate nuclei and unbroken cells from other cell constituents. An aliquot from this nuclear pellet is washed once with TSM buffer and saved for marker enzyme analysis. The supernatant fluid above this pellet is carefully decanted and recentrifuged at 20,000g for 15 min. This gives a pellet composed predominantly of mitochondria, but also contaminated with plasma membrane and endoplasmic reticulum. Aliquots of the mitochondrial pellet are retained in order to determine total recovery of all marker enzymes throughout the procedure. The supernatant largely contains cytosol protein together with plasma membrane and endoplasmic reticulum vesicles. The membrane vesicles can be recovered by centrifugation at 80,000–100,000g for 60 min.

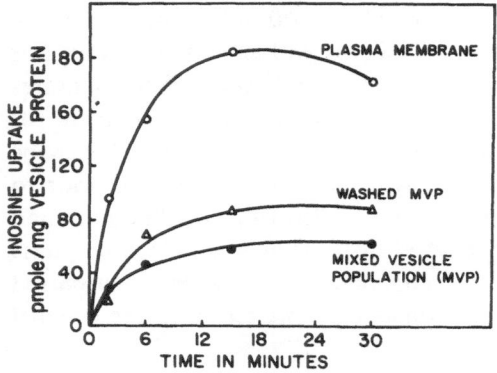

Fig. 1. Inosine transport kinetics for SV-3T3 membrane vesicles at various steps in the plasma membrane isolation procedure. [U-^{14}C]Inosine (254 mCi/mmol) at a final concentration of 50 μM was used. See text for further explanation.

(The latter and all subsequent ultracentrifugations are performed in swinging-bucket rotors.) The resultant pellet, containing both plasma membrane and endoplasmic reticulum material, can be resuspended in TS buffer for vesicles derived from 3T3 and SV-3T3 cells, or TS-2 for BHK vesicles, at about 5 mg membrane vesicle protein per milliliter and stored at −79°C. A 10-ml packed cell volume of SV-3T3 or BHK cells will provide 50–75 mg of protein in the mixed vesicle population. Protein is assayed by the method of Lowry *et al.* (1951) using bovine serum albumin as standard. These membrane vesicles can be used for transport studies, as shown in Figs. 1 and 2.

    e. Recentrifugation of Mitochondrial Material. One modification that has led to significant increases in yield of plasma membrane vesicles from SV-3T3 cells, without loss in transport activity, is recentrifugation of the mitochondrial material (Quinlan and Hochstadt, 1975*a*). As indicated in Table I, the mitochondrial fraction isolated by differential centrifugation

Fig. 2. Transport kinetics for adenosine, inosine, and uridine by isolated plasma membrane vesicles and by isolated endoplasmic reticulum vesicles from SV-3T3 cells. [U-^{14}C]Adenosine (495 mCi/mmol) and [U-^{14}C]uridine (538 mCi/mmol) were used at a final concentration of 30 μM, while [U-^{14}C]inosine (254 mCi/mmol) was used at a final concentration of 50 μM. See text for further details.

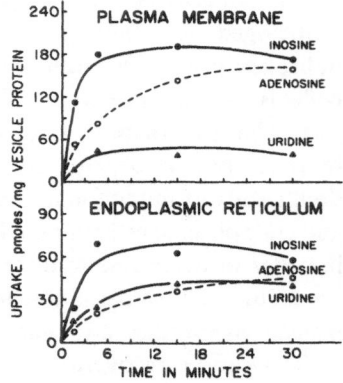

Table I. Distribution of Three Enzyme Activities in Cell Fractions of Balb/c SV-3T3

Cell fractions from SV-3T3[a]	5'-Nucleotidase		NADH dehydrogenase		Succinic dehydrogenase	
	Specific activity[b]	Percent distribution	Specific activity[b]	Percent distribution	Specific activity[b]	Percent distribution
Total homogenate I	0.170	100	63.0	100	41.0	100
Nuclei I	0.042	7.00	82.0	9.00	75.0	25.0
Mitochondria I	1.01	35.0	342	17.0	280	60.0
Mitochondria II	0.720	15.0	243	13.0	302	74.0
Mixed vesicle population II	1.21	78.0	675	48.0	4.00	0.520
Plasma membrane I	4.43	12.0	9.00	0.010	4.00	0.200
Plasma membrane II	4.92	38.0	13.0	0.100	7.00	0.460
Endoplasmic reticulum II	0.682	12.0	703	34.0	11.0	0.500
Percent recovery relative to homogenate[c]		72		56		100

[a] Fractions designated I refer to material obtained by the standard Wallach and Kamat (1966b) fractionation, whereas the designation II refers to material obtained following recentrifugation of mitochondrial fraction I as described in Section 3.3.2e.

[b] nmol/min/mg protein.

[c] The percent relative to the cell homogenate includes values from nuclei, mitochondria II, plasma membrane II, and endoplasmic reticulum.

from SV-3T3 cells contains cosedimenting plasma membranes and endo-plasmic reticulum. The mitochondrial pellet can be resuspended in TSM buffer (this should be done gently so as not to disrupt the mitochondria) and layered over a "cushion" of 16% (w/v) dextran I solution. When centrifuged at 52,000g for 90–120 min, a mitochondrial pellet is obtained, and the vesicles of plasma membrane and endoplasmic reticulum layer on top of the dextran "cushion." This mixed vesicle population can be removed by careful aspiration with a syringe fitted with a needle or cannula, washed once with TSM buffer, centrifuged at 100,000g for 45 min, and pooled with the initial mixed vesicle population described above.

Although this procedure was applied on several occasions to BHK cells, the putative mixed vesicle population recovered from the mitochondrial recentrifugation remained heavily contaminated with mitochondrial material and nucleoside transport activity was lost. The procedure is therefore recommended only for SV-3T3 cells.

f. Separation of Plasma Membranes and Endoplasmic Reticulum from the Mixed Vesicle Population. The mixed vesicle population is diluted with 5–10 vol of 10 mM tris-HCl (pH 8.0) and collected by centrifugation at 100,000g for 40 min. The pellet is suspended in 1 mM tris-HCl (pH 8.0) and the suspension is centrifuged as described above. These hypotonic washing steps are designed to remove inter- and intravesicular "trapped" protein (see Wallach and Kamat, 1966b) and result in a pellet of mixed vesicles that have lost 40–60% of their original protein content.

The pellet is resuspended in 1 mM tris-HCl (pH 8.0) at a protein concentration of about 5 mg/ml and dialyzed at 4°C for 90 min against 200 vol of 1 mM tris-HCl (pH 8.0) containing 1 mM $MgCl_2$ (0.2 mM $MgSO_4$ for BHK vesicles). The dialyzed membrane material is then layered on top of a discontinuous gradient of 10%, 16%, and 23% (w/v) dextran II solutions (dextran III for BHK vesicles). Centrifugation is carried out at 81,500g for 15–18 h for 3T3- and SV-3T3-derived membranes and at 30,000g for 12–13 h for BHK-derived membranes. Two discrete bands form on top of the 10% and 16% dextran II "cushions" which can be removed with a syringe and needle. This plasma membrane fraction is diluted with 5 vol of TS buffer (TS-2 buffer for BHK vesicles). A pellet which also results from this centrifugation step (largely the endoplasmic reticulum fraction) can be removed after the dextran is decanted. The endoplasmic reticulum fraction can be resuspended in TS buffer (or TS-2). Both the endoplasmic reticulum and plasma membrane fractions are washed free of residual dextran using TS (or TS-2) buffer and centrifuged at 80,000–

100,000g for 60 min. The resultant pellets can be resuspended in 0.2-ml aliquots of TS or TS-2 buffer at a concentration of about 5 mg membrane vesicle protein per milliliter and stored frozen at −79°C.

g. Marker Enzyme Analysis of Plasma Membrane and Endoplasmic Reticulum Fractions.

Before describing the marker enzyme analyses used for 3T3, SV-3T3, and BHK cells, it must be stressed again that each cell type may require its own variety of enzymatic or nonenzymatic markers.

Cell and nuclear breakages are estimated as described in Section 3.3.2b. Other subcellular fractions can be analyzed in a more quantitative manner by use of marker enzymes.

Mitochondria are monitored by assaying succinic dehydrogenase activity using a modification of the method of King (1967). The reaction mixture contains, per milliliter, 10 μmol potassium phosphate buffer (pH 7.5), 500 μg bovine serum albumin, 1 μmol potassium cyanide, 10 μmol sodium succinate, 20 μg 2,6-dichlorophenol-indophenol, and, according to enzyme activity, 50–200 μg subcellular protein. The reaction is performed at ambient temperature in a cuvette and the decrease in absorbancy at 600 nm is followed for 5 min. The extinction coefficient of the artificial electron acceptor is 1610 liter cm^{-1} mol^{-1}.

The endoplasmic reticulum is monitored by NADH dehydrogenase activity (Wallach and Kamat, 1966b). The reaction mixture contains, per milliliter, 10 μmol tris-HCl (pH 7.5), 80 μg NADH (Sigma, preweighed at 2 mg/vial), 1.0 μmol potassium ferricyanide, and 5–50 μg subcellular protein. The reaction is performed in a cuvette at ambient temperature, measuring the absorption at 340 nm for 2–3 min. The extinction coefficient for NADH is 6220 liter cm^{-1} mol^{-1}.

5′-Nucleotidase activity, which was used as a marker for the plasma membrane, is assayed by a modification of the method of Weaver and Boyle (1969). The reaction mixture, 50 μl total volume, contains (final concentration) 50 mM tris-HCl (pH 7.5–8.0), 2 mM $MgCl_2$ or $MgSO_4$, 25 μM [U-^{14}C]AMP (≥100 mCi/mmole), and, depending on activity, 1–20 μg subcellular protein. All ingredients except the labeled AMP are preincubated together at 37°C for 5 min prior to the initiation of the reaction by addition of [^{14}C]AMP. Incubation is at 37°C for 10 min. The reaction is stopped by immersing the reaction tubes in ice, adding EDTA to a final concentration of 0.1 M, and boiling for 5 min. A 3- to 4-μl aliquot is then spotted on cellulose thin-layer chromatography sheets containing fluorescent indicator (Eastman Chromagram). Hypoxanthine, adenine, inosine, adenosine, adenylic acid, and inosinic acid serve as chromatographic standards and are prespotted. Development takes place in a solvent composed of butanol–water–propionic acid, 12.5:8.7:6.2 (solvent 2; Hochstadt-Ozer and Stadtman, 1971a). Ultraviolet fluorescence quenching spots are cut out and radioactivity is measured by liquid scintillation spectrometry using toluene plus Liquifluor (New England Nuclear, Boston). It is necessary to follow the recovery of radioactivity in the hypoxanthine, inosine, and adenine "spots ' since these compounds may be derived from adenosine, the product of the 5'-nucleotidase reaction, via action of membrane nucleoside phosphorylase and/or adenosine deaminase (for SV-3T3 preparations, about 50% of the AMP degradation product can be found as inosine and hypoxanthine). The 5'-nucleotidase specific activity is computed in nmol of AMP converted/min/mg protein.

Table I presents marker enzyme data from a typical cell fractionation experiment using Balb/c SV-3T3 cells. Some marker enzymes may be internally localized within selectively permeable organelles and membrane vesicles such that the substrate may not have unrestricted access to the marker enzyme. Therefore, we have often found it necessary to perform duplicate assays using a membrane disaggregating agent such as Triton X-100 (0.05–0.1%, final concentration).

h. Transport Assay. This section will describe procedures used for studying the transport of nucleosides and purine and pyrimidine bases by isolated membrane vesicles from several established mammalian cell lines (Quinlan and Hochstadt, 1974, 1975a; Li and Hochstadt, 1975). The methods are modifications of those procedures reported previously (Hochstadt-Ozer and Stadtman, 1971b,c). Different reaction conditions may be required for each cell type and each transport substrate. For example, the effects of various buffers, ions, and energy sources on transport activity must be empirically determined when membrane vesicles from a new cell line are studied.

The reaction mixture, 100 μl total volume, contains 200 μg membrane vesicle protein, 5.0 μmol potassium phosphate buffer (pH 7.5), 10 μmol sucrose, and radioactively labeled substrate. The concentration of substrate used will, of course, be dependent on the apparent saturability (K_m) of that particular transport system. We have found that concentrations 3- to 4-fold greater than the K_m value result in a significant passive diffusion phenomenon, in addition to the mediated transport process. The reaction mixture, minus substrate, is preincubated 10 min at 37°C. The reaction is initiated by adding the labeled substrate and terminated at various times with 10 vol of 0.8 M NaCl (kept at 37°C). (For BHK membrane vesicles, incubation times in excess of 40 min result in loss of intravesicular transport products, possibly by a "leakage" phenomenon.) After appropriate uptake intervals, the membranes are immediately collected (<20 s) under vacuum on 25- to 30-mm diameter, 0.3-μm pore size nitrocellulose filters. Filters with collected membranes may be rapidly washed once or twice with additional portions of the salt solution used to terminate the reaction. The vacuum is released and filters are removed from the filter apparatus within another 30 s. Thus 30–50 s elapses from reaction termination to removal of filters from the vacuum machine. The filters are dried and counted in a gas flow counter at 17% efficiency (for ^{14}C). Figure 3 shows nucleoside transport kinetics for plasma membrane and endoplasmic reticulum vesicles.

In bacterial membrane studies, it has been determined that terminating the reaction by washing with ice-cold or nonhypertonic solutions leads to significant "leakage" of transported material from the vesicles (Raue and Cashel, 1973; Kaback, 1968; Leder, 1972; Hochstadt, 1974). In accordance with these observations, hypertonic, warm salt washes have been found to be useful in preventing "leakage" of transported material from membrane vesicles prepared from cultured mammalian cells (Hochstadt, 1974; Quinlan and Hochstadt, 1974). For the study of purine nucleoside transport in BHK vesicles, it has been found that inclusion of uridine (10 mM) in the hypertonic wash lowers nonspecific binding of radioactivity to the filters.

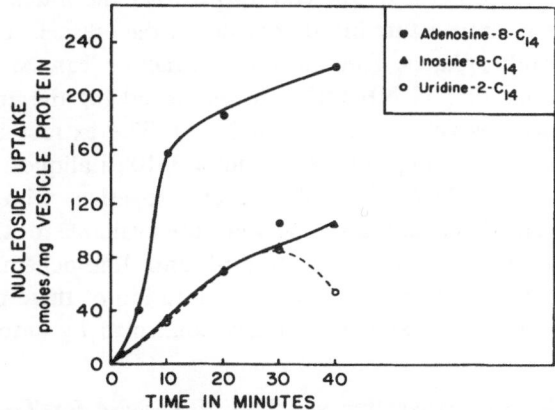

Fig. 3. Transport of [8-^{14}C]adenosine (51 mCi/mmol), [8-^{14}C]inosine (47 mCi/mmol), and [2-^{14}C]uridine (53 mCi/mmol) by a mixed membrane vesicle population prepared from polyoma-transformed BHK cells. The substrate concentration for [8-^{14}C]adenosine and [8-^{14}C]inosine was 100 μM, and for [2-^{14}C]uridine was 88 μM. See text for details of transport assay.

It should be further noted that the transport rate in the mixed vesicle population for an individual nucleoside can fluctuate considerably depending on the particular batch of cells used as a membrane source. For example, using membrane vesicles derived from BHK cells and adenosine as substrate, uptake can vary between 40 and 190 pmol/20 min/mg membrane vesicle protein. This amount of variability has also been reported by Carter *et al.* (1972) and Carter and Martin (1969) for sugar transport by rat adipocytes. Greater consistency seems to be attained when more highly purified plasma membrane vesicles are utilized. Transport kinetics are consistent, though, within each individual batch of membrane vesicles derived from a single cell population.

i. **Chromatographic Analysis of Vesicle Contents.** Transport substrates such as nucleosides and purine and pyrimidine bases may be metabolized before, during, or after the uptake event (Murray *et al.*, 1970). Therefore, it is necessary to identify intravesicular products of transport. Such an examination helps elucidate the possible mechanism of uptake and provides information on the association of metabolic enzymes with the cell membrane.

The following procedure is employed for identifying products of intravesicular transport. The nitrocellulose filters used to collect the membrane vesicles following a transport assay are extracted twice for 15 min with 1 ml of boiling water; more than 85% of the radioactivity is recovered when the two aliquots are pooled. One can also extract the filters with distilled water at ambient temperature in order to detect the presence of any nucleoside di- and triphosphates. Alternatively, extraction can be effected with two 1-ml portions of 2 N NH$_4$OH as is employed for the analysis of the contents of L-cell vesicles (see Section 3.4.5). The extract is lyophilized and resuspended in 30–50 μl of water, and a 3–10 μl aliquot is spotted on cellulose thin-layer chromatography sheets, together with appropriate carriers as described above. Various solvents are available for the separation of nucleic acid precursors (see Schwarz-Mann Radiochemical Catalog, 1970/1971; see also Section 3.3.2g) and the location of these precursors on the thin-layer chromatograms is routinely confirmed by autoradiography.

3.3.3. Discussion of Preparation and Use of Vesicles for Transport

The Wallach and Kamat (1966*b*) nitrogen-cavitation procedure used for preparing membranes has the advantage of producing membrane vesicles, which are both selectively permeable and competent for transport, rather than membranous sheets. However, it has been our experience that the original method gives a plasma membrane yield (based on marker enzyme analysis) of only 10–12%. As shown in Table I, most of the 5'-nucleotidase activity cosediments with the mitochondria (mitochondria fraction I). The use of a 16% dextran barrier to reisolate the mitochondria (mitochondria fraction II) provides additional membrane vesicle material. The data in Table I indicate that this latter material represents about 20% of the total 5'-nucleotidase activity. In addition, there is no succinic dehydrogenase activity contaminating this material. The degree of separation of plasma membrane from endoplasmic reticulum is best examined by comparing the specific activity values of the corresponding marker enzymes. For example, plasma membranes comprise about 10–20% of the endoplasmic reticulum fractions, while 2–5% of the plasma membrane fraction is endoplasmic reticulum. For both fractions, the mitochondrial contamination is less than 2% based on succinic dehydrogenase activity. The plasma membrane fraction contains 35–40% of the 5'-nucleotidase activity, and about 2% of the total protein of the homogenate. The data in Table I show that plasma membrane fraction II contains 3 times more material than plasma membrane I, emphasizing the value of resedimenting mitochondria fraction I to release "trapped" plasma membrane vesicles. These comparisons are, of course, dependent on the validity of the marker enzymes being *exclusively* restricted to a particular cell structure, and being sufficiently active to be detected. Additional marker enzyme analyses that can be utilized in some systems include Na^+/K^+-ATPase and adenyl cyclase for plasma membranes, NADPH-cytochrome c reductase and glucose-6-phosphatase for endoplasmic reticulum, and β-galactosidase and β-glucuronidase for lysosomes (Wallach and Lin, 1973; DePierre and Karnovsky, 1973).

Adenosine, inosine, and uridine uptakes occur by a mediated process in the mixed vesicle population and plasma membrane fraction derived from SV-3T3 cells (Quinlan and Hochstadt, 1974, 1975a, b). One would expect the specific activity of transport by plasma membrane vesicles to increase as the vesicles are further purified if that particular transport system is exclusively associated with the plasma membrane. This would be analogous to enzyme purification. The data in Fig. 1 show that an enrichment of transport activity does occur as plasma membrane vesicles are isolated. For example, the washing steps lead to a 50–60% loss in total protein of the mixed vesicle population while enhancing the specific activity for inosine uptake by about 40%. The dextran-gradient separation of plasma membrane vesicles from the mixed vesicle population leads to a 3-fold enchancement in the ability of plasma membrane vesicles to transport inosine; the concomitant enrichment in $5'$-nucleotidase activity is also about 3- to 4-fold (Table I).

As indicated in Fig. 2, endoplasmic reticulum vesicles may also have transport capacity. Uridine uptake by endoplasmic reticulum vesicles occurs at rates comparable to those observed with plasma membrane vesicles, whereas adenosine transport rates are 4- to 5-fold higher for the plasma membrane vesicles. An alternative explanation of these data would be that the endoplasmic reticulum fraction contains plasma membrane vesicles that have different transport properties than the plasma membrane vesicles recovered from the upper gradient fractions. This would lend support to the concept of a mosaic architecture for surface membranes (Wallach, 1967). A third explanation could be that the plasma membrane vesicles cosedimenting with the endoplasmic reticulum fraction undergo a differential "response" to the fractionation procedure.

Figure 3 shows nucleoside transport by mixed membrane vesicles isolated from polyoma-transformed BHK cells. In the experiments shown, uridine leaked from the vesicles after incubation for 40 min; leakage of all transported nucleosides routinely occurs when incubations are carried out for longer than 40 min. Rates of nucleoside uptake can vary considerably depending on the growth conditions of the cells from which the vesicles are prepared. For example, vesicles from BHK cells grown on DME medium appear to take up adenosine at a higher rate than cells grown on BHK medium. Conversely, vesicles derived from cells grown on BHK medium appear to transport inosine at a higher rate than vesicles isolated from DME grown cells. These patterns, although not conclusive because of variations from batch to batch, suggest that several growth media should be investigated in developing a new substrate transport system. The independent variation of inosine and adenosine transport indicates that uptake of these two purine nucleosides is mediated by distinct transport systems. Also, since the transport substrates used were specifically labeled in the base moiety, the transport product(s) could be the intact nucleoside and/or the free base.

On the assumption that the plasma membrane accounts for approximately 4–5% of total cell protein, the initial rates of transport in isolated vesicles indicate that we recover in plasma about 10–15% of whole cell activity (correcting for recovery of membrane). We presume that the remainder of the transport activity is either destroyed or lost during preparation or inhibited by accumulation of products within the vesicles. This latter possibility could arise if products were not removed by further metabolism, as occurs in the intact cell.

3.4. Preparation of Nucleoside Transport-Active Membrane Vesicles from L-Cell Fibroblasts Grown in Culture

3.4.1. Introduction

The basic principles of the method used for the separation of plasma membranes from other subcellular fractions in L-cell fibroblasts are based on the procedure developed by McKeel and Jarrett (1970). These authors prepared a fat cell plasma membrane fraction in order to study membrane–hormone interactions, while our purpose was to obtain purified plasma membrane vesicles from these cultured mammalian cells, which still retain transport activity. Consequently, the following modifications were introduced: (1) the pH of the buffer used during cell fractionation is 8.0 instead of 7.4 (the higher pH has been found to facilitate cell disruption and may reduce membrane aggregation), (2) the concentration of tris-EDTA is reduced to 0.1 mM (lowering the EDTA concentration has been found to reduce substantially mitochondrial contamination), (3) prior to fractionation the cell homogenate is adjusted to 0.1 mM $MgSO_4$ to minimize disintegration of other subcellular organelles, (4) a cushion of 5% ficoll solution is layered between the sample suspension and the gradient itself to facilitate separation of different plasma membrane subfractions, and (5) the time used for the gradient centrifugation (5 h) is much longer than that used in the original procedure (0.5 h). This modification, although prolonging the time required for plasma membrane isolation, helps reduce mitochondrial contamination. A flowsheet for the procedure is presented in Fig. 4.

3.4.2. Procedure

a. **Cell Culture.** The method described above has been applied to the preparation of plasma membrane vesicles from the following cell lines: (1) recloned L cells (NCTC clone 929 mouse cells); (2) recloned A_9 cells, a hypoxanthine-guanine phosphoribosyltransferase negative, adenine phosphoribosyltransferase negative mutant line selected from L cells by Little-field (1964); (3) recloned A_{9R} cells, and A_9 revertant, which is an adenine phosphoribosyltransferase negative line that is hypoxanthine-guanine phosphoribosyltransferase positive (McBride and Ozer, 1973); (4) L_{929se-} cells, a subline of L cells which have been adapted to grow in completely defined medium; and (5) recloned Chinese hamster ovary (CHO) cells. L_{929se-} cells are grown in roller bottles (720 cm^2) containing 150 ml Weymouth medium MD 705 per liter (Grand Island Biological Company) without addition

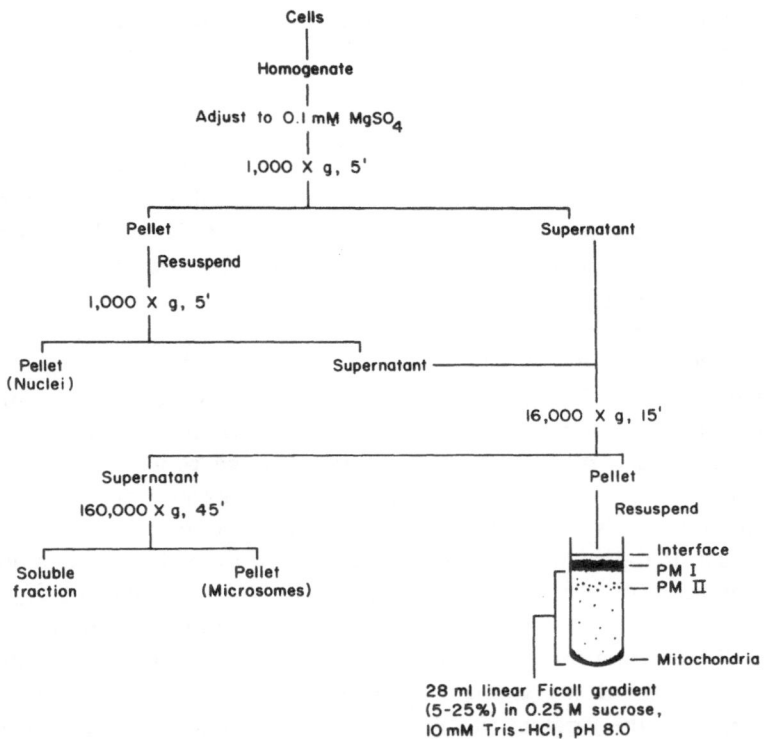

Fig. 4. Flowsheet for the isolation of L-cell plasma membrane vesicles. Cells were grown, harvested, and fractionated as described in the text.

of serum or any macromolecular preparation. At confluence, each bottle yields approximately 2.8×10^5 cells/cm². All other cell lines are grown in suspension culture in Joklik modified Eagle's minimal essential medium supplemented with 10% fetal calf serum; cell densities are maintained at $2–3 \times 10^5$ cells per milliliter.

b. Cell Washing and Homogenization. All cell washing and homogenization steps are carried out at 4°C. Cells are harvested by scraping and are washed three times in 50-vol portions of 0.5 M sucrose, 10 mM tris-HCl (pH 8.0), and 0.1 mM EDTA-tris (pH 8.0) (medium I). Centrifugation is performed at 800g for 10 min. The washed cells are resuspended in 20 vol of medium I and kept at 2°C for 0.5 h to facilitate cell disruption. Homogenization consists of 2–5 strokes at 1800 rpm with a glass homogenizer fitted with a motor-driven teflon pestle. Cell breakage is monitored by phase contrast microscopy. The extent of nuclear rupture is estimated by comparing the cell number before homogenization to nuclei and cell numbers after homogenization. In general, greater than 92% cell breakage can be obtained without causing significant nuclear breakage (less than 2%).

    c. Removal of the Nuclear and Microsomal Fractions from the Homogenate. The homogenate is adjusted to 0.1 mM $MgSO_4$ and nuclei and unbroken cells are removed by centrifugation at 1000g for 5 min; the supernatant is decanted and saved. The pellet is resuspended in 10 vol of medium I containing 0.1 mM $MgSO_4$ using a Dounce homogenizer and centrifuged as before. The two supernatant fluids are pooled and centrifuged at 16,000g for 15 min. The resulting pellet is to be used for the preparation of the plasma membrane fraction as described below. The supernatant fluid is centrifuged at 160,000g for 45 min in order to sediment a microsomal fraction, which contains predominantly endoplasmic reticulum and ribosomes.

    d. Separation of Plasma Membranes and Mitochondria. The pellet obtained from the 15-min 16,000g centrifugation step described above is resuspended in 4–6 vol (based on original packed cell volume) of medium I. A 28-ml linear gradient is formed from 5 to 25% ficoll in medium I. Immediately before use, 3 ml of the 5% ficoll solution is layered on top of the gradient followed by 7 ml of sample suspension. Centrifugation is performed using a SW27 rotor at 56,000g for 5 h. Plasma membrane vesicles form two visible bands in the gradient; one is a compact band just below the sample–ficoll interface and the other is a diffuse, faint-colored area about a third of the way into the gradient. These two bands are collected separately and diluted, respectively, 4:1 and 6:1 (v/v) with medium I, and centrifuged at 16,000g for 20 min. The pellets are resuspended in 0.25 M sucrose, 10 mM tris-HCl (pH 8.0) (medium II), and labeled PM I and PM II, respectively. When the gradient centrifugation is performed for shorter times (e.g., 56,000g for 45 min), the PM II material forms a more distinct band below the PM I band. However, PM II is then rather heavily contaminated by mitochondria. The firm pellet at the bottom of the tube is composed of mitochondria. The latter (as well as the microsomes and nuclei) are resuspended in medium II for storage at −79°C.

3.4.3. Comments on the Procedure

    a. General. The procedure, as described above, requires 9–15 h. Alternatively, the gradient centrifugation time can be extended to 17–18 h at 33,000g. Thus centrifugation may be started in the late afternoon and membrane fractions collected the following morning. The option of extending the centrifugation overnight is more convenient and does not affect enzyme and transport activity. The reproducibility of the procedure is excellent with respect to marker enzymes.

Although the maximum velocities (V_{max}) for nucleoside transport may vary from one vesicle preparation to another, this variability again appears to be due to the state of growth of the cells from which the membranes are derived. In general, plasma membranes prepared from actively growing cells have higher transport rates than membranes prepared from confluent cultures. The substrate saturation curve constants (K_m) for the transport systems are, however, essentially constant.

b. Purity and Yield. The purity and yield of plasma membranes are estimated by the use of two marker enzymes: Na^+/K^+-ATPase and 5'-AMPase. In addition, NADH dehydrogenase and succinate dehydrogenase were selected as marker enzymes for the microsomes and mitochondria, respectively. The enzyme assays are performed according to the procedures described by Avruch and Wallach (1971) for Na^+/K^+-ATPase, 5'-AMPase, and NADH dehydrogenase. King's method (1967) is used for succinic dehydrogenase (see Section 3.3.2g for assay protocols). Table II shows the distribution and enrichment of these enzymes in various subcellular fractions obtained from L cells. It is interesting to note that the two plasma

Table II. Enrichment of Subcellular Marker Enzymes in Various Fractions of L Cells[a]

Fraction	Na^+/K^+-ATPase	5'-AMPase	NADH dehydrogenase	Succinic dehydrogenase
Homogenate	0.84 (100)	1.15 (100)	13.1 (100)	15.3 (100)
Nuclei	0.61 (10)	0.5 (6.6)	0.34 (1.0)	11.9 (10.8)
Plasma membrane I	18.8 (38)	2.45 (4.0)	1.7 (0.6)	8.9 (1.0)
Plasma membrane II	6.0 (6.6)	16.9 (18)	4.4 (1)	11.2 (0.9)
Mitochondria	0.93 (6.0)	2.78 (16)	0.75 (1.0)	154.8 (65)
Microsomes	0.80 (12)	0.93 (10)	56.4 (55)	10.0 (8.4)

[a] Specific activity: nmol converted/min/mg protein. The values in parentheses are percent distributions of enzyme activity relative to the homogenate.

membrane marker enzymes do not cofractionate with respect to plasma membrane fractions I and II; I is enriched for Na+/K+-ATPase and II for 5'-AMPase. The densities of these two plasma membrane fractions, as measured by weighing a known volume of the material taken from the center of each band, are 1.054 ± 0.006 g/cm³ (fraction I) and 1.073 ± 0.006 g/cm³ (fraction II). Significant levels of the plasma membrane marker enzyme activities are seen in mitochondria (5'-AMPase) and in microsomes (Na+/K+-ATPase). However, contamination of the plasma membrane preparation by mitochondria and microsomes appears to be minimal. The yield of plasma membrane is relatively low, as shown in Table II, but, as has been pointed out by Wallach and Lin (1973), the membrane enzymes may be eluted or inactivated during fractionation and therefore estimation of membrane yield by the recovery of marker enzymes may not be accurate. The yield of total cell protein in the plasma membrane fractions from different sublines of L cells is in the range of 1.6–2.6% of total homogenate protein.

3.4.4. Assay for Nucleoside Transport in Vesicles Prepared from L-Cell Fibroblast Lines

Assays for nucleoside transport were conducted as described for the 3T3 and BHK cell lines (Section 3.3.2h), with the following exceptions. Membrane vesicles derived from certain L-cell sublines show consistently higher nucleoside transport rates than others (Li and Hochstadt, 1975), so that as little as 40 μg membrane protein could be added to each reaction mixture. A 5-min preincubation, rather than 10 min, at 37°C was found to be optimal for vesicles from L cells; 0.1 M NaCl replaced 0.1 M sucrose as an osmotic stabilizer in the reaction mixture.

3.4.5. Analysis of Transport Products

Vesicle contents are determined by chromatography and autoradiography as described in Section 3.3.2i, with the exception that 2 N NH₄OH is always employed to extract the filters, and additional solvents have been found useful in the separation of the nucleic acid precursors. For example, saturated $(NH_4)_2SO_4$–0.1 M potassium phosphate (pH 6.0)–isopropanol, 79:19:2, produces the following migration pattern from the origin to the solvent front:bases, nucleosides, nucleotides. This solvent system is therefore useful in the initial determination of whether any metabolism of the substrate has occurred.

Table III. Rates of Nucleoside Transport by Membrane Vesicles from Different
Cell Sources[a]

| Fraction | L cells | | CHO cells |
	Adenosine	Inosine	Adenosine
Plasma membrane I	11.7	25.5	52
Plasma membrane II	23.3	12.4	42.5
Mitochondria	1.95	3.05	17.5
Microsomes	0	6.5	5.0

[a] Transport rates expressed as pmol/mg protein/min.

3.4.6. Conclusions Concerning L-Cell Vesicle Preparation and Transport Assay

Plasma membrane fractions I and II both transport nucleosides, although at different initial rates. The transport activities of the mitochondrial and microsomal fractions, while not insignificant, are substantially lower than the plasma membrane preparations and could be attributable to contamination by plasma membranes. Initial rates for adenosine and inosine transport, as measured under the standard conditions described above, are listed in Table III. The transport activities of plasma membrane fractions isolated from L cells and CHO cells are also compared in Table III. Plasma membrane derived from CHO cells has V_{max} values at least 2-fold higher than plasma membrane from L cells.

We conclude that a modified McKeel and Jarrett (1970) method can also be employed to generate plasma membrane vesicles capable of transporting nucleosides by a mediated uptake process (Li and Hochstadt, 1975). This method, together with the nitrogen-cavitation procedure, should make it possible to broaden the scope of uptake studies in terms of transport substrates used and in the use of additional cultured, mammalian cell lines. It must be stressed that each cell line and transport system examined may require modifications in the cell fractionation and/or transport assays.

4. CONCLUSIONS

Considerable insight and information have been gained from experimentation with isolated membranes. Bacterial vesicles have proved of increasing value in elucidating transport mechanisms such as group translocation as well as in unraveling the manner in which energy is coupled to transport. The experience and techniques which have been developed with bacterial vesicles are presently being applied to mammalian systems. These studies may contribute to an understanding of the role of transport function in growth control.

While the importance of vesicles for transport studies is becoming more evident, the methodologies for vesicle preparation remain rather limited. The preparation of bacterial membranes involves the conversion of the bacteria to osmotically sensitive forms which are then lysed to yield closed vesicles. Most investigators have relied on the Kaback technique, but we have found an alternative method that is often advantageous for the consistent production of membrane vesicles from numerous strains of *Salmonella typhimurium* that will transport a variety of substrates. This procedure may prove useful for other bacterial species.

The techniques for isolating mammalian membranes are more varied and can be divided into two areas: cell disruption and purification of membranes. Presently, Dounce homogenization and nitrogen cavitation are popular disruption techniques that will yield membranes suitable for transport studies. The problem of preventing simultaneous lysis of intracellular organelles still exists. Endoplasmic reticulum is a common contaminant of the plasma membrane fraction when the nitrogen-cavitation procedure is employed, but this problem is largely resolved with the introduction of ficoll and dextran gradients. Mitochondria, on the other hand, are a minor contaminant of plasma membranes prepared by the Dounce homogenization procedure. Comparison of results obtained using the two procedures should allow one to rule out effects that are due to either contaminant. The use of isolation techniques involving affinity columns, which could selectively retain various membrane fractions, would have the advantages of providing membranes of greater purity in higher yields and in less time than the present procedures require. While improvements in membrane isolation techniques are needed, the currently available methodologies yield membranes which can be studied in order to enhance our understanding of transport mechanisms.

The emphasis in this chapter has been on methodology, for the purpose of showing how a variety of techniques already available can be employed. For the bacterial system, it is of interest to report in detail on alternative uses of a well-established procedure, as well as to suggest alternative techniques. Several procedures resulting in isolated membrane vesicles for transport studies, the rationale for employing them, and the results which have been obtained have been discussed in detail. Progress in the area of group translocation and the mechanism of energy coupling to transport have been cited as examples of how useful bacterial membrane vesicle work has been in the generation of new concepts in transport and related biochemical reactions occurring at the cell surface. The need for new isolation techniques for the study of transport seems apparent. We have presented an adaption of a procedure based on the isolation technique of Obsorn *et al.* (1972).

Because mammalian cells have become the focus for studies relating to growth control mechanisms as well as how transport may be involved in such control, we have developed an experimental system for the study of transport in membrane vesicles isolated from cultured mammalian cells. We have emphasized uptake of nucleic acid precursors although we envision that these techniques will be applied to the study of uptake mechanisms for a variety of substrates and cell lines. Finally, better understanding of transport function in animal cells may contribute significantly to the elucidation of overall membrane regulatory and cell regulatory mechanisms.

ACKNOWLEDGMENTS

Work described in this chapter was supported by USPHS grants CA 14780 and GM 20486 and an Established Investigatorship of the American Heart Association to J. H. We also wish to thank Drs. Harvey Ozer, Grant Fairbanks, and Robert Weihing for their constructive criticism during the preparation of the manuscript.

NOTE ADDED IN PROOF

For preparation of plasma membrane vesicles from new cell lines, we suggest the mixed vesicle population be prepared as described on pp. 138–142. If the dextran step does not yield well separated PM and ER by marker analysis as described, then continuous dextran gradients should be prepared at several Mg^{2+} concentrations between 0.05 mM and 2mM. The Mg^{2+} concentration giving well separated PM and ER material can then be employed using a dextran concentration of density midway between PM and ER isopyanic banding.

5. REFERENCES

Allan, D., and Crumpton, M. J., 1970, Preparation and characterization of the plasma membrane of pig lymphocytes, *Biochem. J.* **120**:133.

Altendorf, K. H., and Staehelin, L. A., 1974, Orientation of membrane vesicles from *Escherichia coli* as detected by freeze-cleave electron microscopy, *J. Bacteriol.* **117**:888.

Avruch, J., and Wallach, D. F. H., 1971, Preparation and properties of plasma membrane and endoplasmic reticulum fragments from isolated rat fat cells, *Biochim. Biophys. Acta* **233**:334.

Barnes, E. M., Jr., 1972, Respiration-coupled glucose transport in membrane vesicles from *Azotobacter vinelandii*, *Arch. Biochem.* **152**:795.

Barnes, E. M., Jr., 1973, Multiple sites for coupling of glucose transport to the respiratory chain of membrane vesicles from *Azotobacter vinelandii*, *J. Biol. Chem.* **248**:8120.

Barnes, E., Jr., 1974, Respiration-coupled calcium transport by membrane vesicles from *Azotobacter vinelandii*, *Fed. Proc.* **33**:1457.

Barnes, E. M., Jr., and Kaback, H. R., 1970, β-Galactoside transport in bacterial membrane preparations: Energy coupling via membrane-bound D-lactic dehydrogenase, *Proc. Natl. Acad. Sci. (USA)* **66**:1190.

Benke, P. J., Herrick, N., and Hebert, A., 1973, Transport of hypoxanthine in fibroblasts with normal and mutant hypoxanthine-guanine phosphoribosyltransferase, *Biochem. Med.* **8**:309.

Berger, E. A., 1973, Different mechanisms of energy coupling for the active transport of proline and glutamine in *Escherichia coli*, *Proc. Natl. Acad. Sci. (USA)* **70**:1514.

Bhattacharyya, P., 1970, Active transport of manganese in isolated membranes of *Escherichia coli*, *J. Bacteriol.* **104**:1307.

Bhattacharyya, P., Epstein, W., and Silver, S., 1971, Valinomycin-induced uptake of potassium in membrane vesicles from *Escherichia coli*, *Proc. Natl. Acad. Sci. (USA)* **68**:1488.

Brunette, D. M., and Till, J. E., 1971, A rapid method for the isolation of L-cell surface membrane using an aqueous two-phase polymer system, *J. Membr. Biol.* **5**:215.

Carter, J. E., and Martin, D. B., 1969, Glucose uptake by isolated particles from rat epididymal adipose tissue cells, *Proc. Natl. Acad. Sci. (USA)* **64**:1343.

Carter, J. E., Avruch, J., and Martin, D. B., 1972, Glucose transport in plasma membrane vesicles from rat adipose tissue, *J. Biol. Chem.* **246**:2682.

Cunningham, D. D., and Pardee, A. B., 1969, Transport changes rapidly initiated by serum addition to "contact-inhibited" 3T3 cells, *Proc. Natl. Acad. Sci. (USA)* **64**:1049.

DePierre, J. W., and Karnovsky, M. L., 1973, Plasma membranes of mammalian cells: A review of methods for their characterization and isolation, *J. Cell Biol.* **56**:257.

Frerman, F. E., and Bennett, W., 1973, Studies on the uptake of fatty acids by *Escherichia coli*, *Arch. Biochem. Biophys.* **159**:434.

Futai, M., 1974*a*, Reconstitution of transport dependent on D-lactate or glycerol-3-phosphate in membrane vesicles of *Escherichia coli* deficient in the corresponding dehydrogenases, *Biochemistry* **13**:2327.

Futai, M., 1974*b*, Orientation of membrane vesicles from *Escherichia coli* prepared by different procedures, *J. Membr. Biol.* **15**:15.

Gahmberg, C. G., and Simons, K., 1970, Isolation of plasma membrane fragments from BHK 21 cells, *Acta Pathol. Microbiol. Scand. Sect. B* **78**:176.

Graham, J., 1972, Isolation and characterization of membranes from normal and transformed tissue culture cells, *Biochem. J.* **130**:1113.

Graham, J. M., Sumner, M. C. B., Curtis, D. H., and Pasternek, C. A., 1973, Sequence of events in plasma membrane assembly during the cell cycle, *Nature* **246**:291.

Gruenstein, E., Rich, A., and Weihing, R. R., 1974, Actin associated with membranes from 3T3 mouse fibroblasts and HeLa cells, *J. Cell Biol.*, **64**:223.

Hampton, M. L., and Freese, E., 1974, Explanation for the apparent inefficiency of reduced nicotinamide adenine dinucleotide in energizing amino acid transport in membrane vesicles, *J. Bacteriol.* **118**:497.

Harold, F. M., 1972, Conservation and trasformation of energy of bacterial membranes, *Bacteriol. Rev.* **36**:172.

Heppel, L. A., 1971, The concept of periplasmic enzymes, in: *Structure and Function of Biological Membranes* (L. Rothfield, ed.), p. 223, Academic Press, New York.

Hinds, T. R., and Brodie, A. F., 1974, Relationship of a proton gradient to the active transport of proline with membrane vesicles from *Mycobacterium phlei*, *Proc. Natl. Acad. Sci. (USA)* **71**:1202.

Hirata, H., Asano, A., and Brodie, A. F., 1971, Respiration dependent transport of proline by electron transport particles from *Mycobacterium phlei*, *Biochem. Biophys. Res. Commun.* **44**:368.

Hirata, H., Altendorf, K., and Harold, F. M., 1974, Energy coupling in membrane vesicles of *Escherichia coli*. I. Accumulation of metabolites in response to an electrical potential, *J. Biol. Chem.* **249**:2939.

Hochstadt, J., 1974, The role of the membrane in the utilization of nucleic acid precursors, *CRC Crit. Rev. Biochem.* **2**:259.

Hochstadt-Ozer, J., 1972, The regulation of purine utilization in bacteria. IV. Roles of membrane-localized and pericytoplasmic enzymes in the mechanism of purine nucleoside transport across isolated *E. coli* membranes, *J. Biol. Chem.* **247**:2419.

Hochstadt-Ozer, J., and Cashel, M., 1972, The regulation of purine utilization in bacteria. V. Inhibition of purine phosphoribosyltransferase activity and purine uptake in isolated membrane vesicles by guanosine tetraphosphate, *J. Biol. Chem.* **247**:7067.

Hochstadt-Ozer, J., and Stadtman, E. R., 1971*a*, The regulation of purine utilization in bacteria. I. Purification of adenine phosphoribosyltransferase and control of activity by nucleotides, *J. Biol. Chem.* **246**:5294.

Hochstadt-Ozer, J., and Stadtman, E. R., 1971*b*, The regulation of purine utilization in bacteria. II. Adenine phosphoribosyltransferase in isolated membrane preparations and its role in transport of adenine across the membrane, *J. Biol. Chem.* **246**:5304.

Hochstadt-Ozer, J., and Stadtman, E. R., 1971*c*, The regulation of purine utilization in bacteria. III. The involvement of purine phosphoribosyltransferase in the uptake of adenine and other nucleic acid precursors by intact resting cells, *J. Biol. Chem.* **246**:5312.

Holley, R. W., 1972, A unifying hypothesis concerning the nature of malignant growth, *Proc. Natl. Acad. Sci. (USA)* **69**:2840.

Kaback, H. R., 1968, The role of the phosphoenolpyruvate phosphotransferase system in the transport of sugars by isolated membrane preparations of *Escherichia coli*, *J. Biol. Chem.* **243**:3711.

Kaback, H. R., 1970, Transport, *Ann. Rev. Biochem.* **39**:561.

Kaback, H. R., 1971, Bacterial membranes, in: *Methods in Enzymology*, Vol. 22 (W. B. Jacoby, ed.), p. 99, Academic Press, New York.

Kaback, H. R., 1972, Transport across isolated bacterial cytoplasmic membranes, *Biochim. Biophys. Acta* **265**:367.

Kaback, H. R., and Hong, J., 1973, Membranes and transport, *CRC Crit. Rev. Microbiol.* **2**:333.

Kaback, H. R., and Milner, L. S., 1970, Relationship of a membrane-bound D-lactic dehydrogenase to amino acid transport in isolated bacterial membrane preparations, *Proc. Natl. Acad. Sci. (USA)* **66**:1008.

Kaback, H. R., and Stadtman, E. R., 1966, Proline uptake by an isolated cytoplasmic membrane preparation of *Escherichia coli*, *Proc. Natl. Acad. Sci. (USA)* **55**:920.

Kaback, H. R., and Stadtman, E. R., 1968, Glycine uptake in *Escherichia coli*. II. Glycine uptake, exchange, and metabolism by an isolated membrane preparation, *J. Biol. Chem.* **243**:1390.

Kamat, V. B., and Wallach, D. F. H., 1965, Separation and partial purification of plasma-membrane fragments from Ehrlich ascites carcinoma microsomes, *Science* **148**:1343.

Kashket, E. R., and Wilson, T. H., 1973, Proton-coupled accumulation of galactoside in *Streptococcus lactis* 7962, *Proc. Natl. Acad. Sci. (USA)* **70**:2866.

Kerwar, G., Gordon, A. S., and Kaback, H. R., 1972, Mechanisms of active transport in isolated membrane vesicles. IV. Galactose transport by isolated membrane vesicles from *Escherichia coli*, *J. Biol. Chem.* **247**:291.

King, T. E., 1967, Preparations of succinate-cytochrome *c* reductase and the cytochrome *b*-*c* particle, and reconstitution of succinate-cytochrome *c* reductase, in: *Methods in Enzymology*, Vol. 10 (R. W. Estabrook and M. E. Pullman, eds.), p. 216, Academic Press, New York.

Klein, W. L., and Boyer, P. D., 1972, Energization of active transport by *Escherichia coli*, *J. Biol. Chem.* **247**:7257.

Komatsu, Y., and Tanaka, K., 1973, Deoxycytidine uptake by isolated membrane vesicles from *Escherichia coli* K_{12}, *Biochim. Biophys. Acta* **311**:496.

Konings, W., and Freese, E., 1972, Amino acid transport in membrane vesicles of *Bacillus subtilis*, *J. Biol. Chem.* **247**:2408.

Konings, W., and Kaback, H. R., 1973, Anaerobic transport in *Escherichia coli* membrane vesicles, *Proc. Natl. Acad. Sci. (USA)* **70**:3376.

Konings, W. N., Barnes, E. M., Jr., and Kaback, H. R., 1971, Mechanisms of active transport in isolated membrane vesicles. III. The coupling of reduced phenazine methosulfate to the concentrative uptake of β-galactosides and amino acids, *J. Biol. Chem.* **246**:5857.

Konings, W. N., Bisschop, A., Veenhuis, M., and Vermeulen, C. A., 1973, New procedure for the isolation of membrane vesicles of *Bacillus subtilis* and an electron microscopy study of their ultrastructure, *J. Bacteriol.* **116**:1456.

Lauter, C. J., Solyom, A., and Trams, E. G., 1972, Comparative studies on enzyme markers of liver plasma membranes, *Biochim. Biophys. Acta* **266**:511.

Leder, I., 1972, Interrelated effects of cold shock and osmotic pressure on the permeability of the *E. coli* membrane to permease accumulated substrates, *J. Bacteriol.* **111**:211.

Lesko, L., Donlon, M., Marinetti, G. V., and Hare, J. D., 1973, A rapid method for the isolation of rat liver plasma membranes using an aqueous two-phase polymer system, *Biochim. Biophys. Acta* **311**:173.

Li, C. C., and Hochstadt, J., 1975, Nucleoside uptake by plasma membrane vesicles from L929 cells grown in completely defined medium, submitted for publication.

Littlefield, J. W., 1964, Three degrees of guanylic acid-inosinic acid pyrophosphorylase deficiency in mouse fibroblasts, *Nature* **203**:1142.

Lombardi, F. J., and Kaback, H. R., 1972, Mechanisms of active transport in isolated bacterial membrane vesicles. VIII. The transport of amino acids by membranes prepared from *Escherichia coli*, *J. Biol. Chem.* **247**:7844.

Lombardi, F. J., Reeves, J. P., and Kaback, H. R., 1973, Mechanisms of active transport in isolated bacterial membrane vesicles. XIII. Valinomycin-induced rubidium transport, *J. Biol. Chem.* **248**:3551.

Lowry, O. H., Roseborough, N. J., Farr, A. J., and Randell, R. J., 1951, Protein measurement with the Folin phenol reagent, *J. Biol. Chem.* **193**:265.

MacLeod, R. A., Thurman, P., and Rogers, H. J., 1973, Comparative transport activity of intact cells, membrane vesicles, and mesosomes of *Bacillus licheniformis*, *J. Bacteriol.* **113**:329.

McBride, O. W., and Ozer, H. L., 1973, Transfer of genetic information by purified metaphase chromosomes, *Proc. Natl. Acad. Sci. (USA)* **70**:1258.

McKeel, D. W., and Jarrett, L., 1970, Preparation and characterization of a plasma membrane fraction from isolated fat cells, *J. Cell Biol.* **44**:417.

Meezan, E., Wu, H., Black, P. A., and Robbins, P. W., 1969, Comparative studies on the carbohydrate-containing membrane components of normal and virus-transformed mouse fibroblasts. II. Separation of glycoproteins and glycopeptides by Sephadex chromatography, *Biochemistry* **8**:2518.

Murray, A. W., Elliott, D. C., and Atkinson, M. R., 1970, Nucleotide biosynthesis from preformed purines in mammalian cells: Regulatory mechanisms and biological significance, in: *Progress in Nucleic Acid Research and Molecular Biology*, Vol. 10 (J. N. Davidson and W. E. Cohn, eds.), p. 87, Academic Press, New York.

Neville, D. M., 1960, The isolation of a cell membrane fraction from rat liver, *J. Biophys. Biochem. Cytol.* **8**:413.

Neville, D. M., 1975, Isolation of cell surface membrane fractions from mammalian cells and organs, in: *Methods in Membrane Biology*, Vol. 3 (E. D. Korn, ed.), p. 1, Plenum Press, New York.

Noonan, K., Levine, A., and Burger, N., 1973, Cell cycle–dependent changes in the surface membrane as detected with ^{3}H-concanavalin A, *J. Cell Biol.* **58**:491.

Osborn, M. J., Gander, J. E., Parisi, E., and Carson, J., 1972, Mechanism of assembly of the outer membrane of *Salmonella typhimurium*: Isolation and characterization of cytoplasmic and outer membrane, *J. Biol. Chem.* **247**:3962.

Ozer, J. (Hochstadt), and Wallach, D. F. H., 1967, *H-2* components and cellular membranes: Distinctions between plasma membrane and endoplasmic reticulum governed by the *H-2* region in the mouse, *Transplantation* **5**:652.

Perdue, J. F., 1971, The isolation and characterization of plasma membranes from cultured cells. III. The adenosine triphosphate-dependent accumulation of Ca^{2+} by chick embryo fibroblasts, *J. Biol. Chem.* **246**:6750.

Pickard, M. A., Phillippe, L., and Campbell, J. N., 1974, Metabolism and transport of purine nucleosides by membrane preparations of *Micrococcus sodonensis*, *Can. J. Biochem.* **52**:83.

Plagemann, P. G. W., and Erbe, J., 1972, Thymidine transport by cultured Novikoff hepatoma cells and uptake by simple diffusion and relationship to incorporation in deoxyribonucleic acid, *J. Cell Biol.* **55**:161.

Plagemann, P. G. W., and Erbe, J., 1973, Nucleotide pools in Novikoff rat hepatoma cells growing in suspension culture. IV. Nucleoside transport in cells depleted of nucleotides by treatment with KCN, *J. Cell Physiol.* **81**:101.

Post, R. L., and Jolly, P. C., 1957, The linkage of Na^+, K^+ and NH_4^+ active transport across the human erythrocyte membrane, *Biochim. Biophys. Acta* **25**:118.

Prezioso, G., Hong, J., Kerwar, G. N., and Kaback, H. R., 1973, Mechanisms of active transport in isolated bacterial membrane vesicles. XII. Active transport by a mutant of *Escherichia coli* uncoupled for oxidative phosphorylation, *Arch. Biochem. Biophys.* **154**:575.

Quinlan, D. C., and Hochstadt, J., 1974, An altered rate of uridine transport in membrane vesicles isolated from growing and quiescent 3T3 cells, *Proc. Natl. Acad. Sci. (USA)* **71**:5000.

Quinlan, D. C., and Hochstadt, J., 1975a, Mechanisms of transport by isolate membranes from culture mammalian cells: Group translocation of the ribose moiety of inosine by plasma membrane vesicles from 3T3 cells transformed with Simian virus 40, *J. Biol. Chem.*, in press.

Quinlan, D. C., and Hochstadt, J., 1975b, The existence of a group translocation mechanism in animal cells: Uptake of the ribose moiety of inosine, *J. Supramolecular Structure*, in press.

Raue, H. A., and Cashel, M., 1973, Regulation of RNA synthesis in *E. Coli.* I. Characterization of cells subjected to simultaneous temperature and osmotic shock, *Biochim. Biophys. Acta* **312**:722.

Ray, T. K., 1970, A modified method for the isolation of the plasma membrane from rat liver, *Biochim. Biophys. Acta* **196**:1.

Reeves, J. P., 1971, Transient pH changes during D-lactate oxidation by membrane vesicles, *Biochem. Biophys. Res. Commun.* **45**:931.

Reeves, J. P., Hong, J., and Kaback, H. R., 1973, Reconstitution of D-lactate dependent transport in membrane vesicles from a D-lactate dehydrogenase mutant of *Escherichia coli*, *Proc. Natl. Acad. Sci. (USA)* **70**:1917.

Romano, A. H., and Colby, C., 1973, SV40 virus transformation of mouse 3T3 cells does not specifically enhance sugar transport, *Science* **179**:1238.

Roseman, S., 1969, The transport of carbohydrates by a bacterial phosphotransferase system, *J. Gen. Physiol.* **54**:1385.

Rosen, B. P., 1973, Restoration of active transport in a Mg^{2+}-adenosine triphosphatase-deficient mutant of *Escherichia coli*, *J. Bacteriol.* **116**:1124.

Schuster, G. S., and Hare, J. D., 1971, The role of phosphorylation in the uptake of thymidine in mammalian cells, *In Vitro* **6**:427.

Shapiro, B. M., Sicardi, A. G., Hirota, Y., and Jacob, F., 1970, On the process of cellular division in *Escherichia coli*. II. Membrane protein alterations associated with mutations affecting the initiation of DNA synthesis, *J. Mol. Biol.* **52**:75.

Short, S. A., White, D. C., and Kaback, H. R., 1972a, Mechanisms of active transport in isolated bacterial membrane vesicles. IX. The kinetics and specificity of amino acid transport in *Staphylococcus aureus* membrane vesicles, *J. Biol. Chem.* **247**:7452.

Short, S. A., White, D. C., and Kaback, H. R., 1972b, Active transport in isolated bacterial membrane vesicles. V. The transport of amino acids by membrane vesicles prepared from *Staphylococcus aureus*, *J. Biol. Chem.* **247**:298.

Short, S. A., Kaback, H. R., and Kohn, L. S., 1974, D-Lactate dehydrogenase binding in *Escherichia coli* dld-membrane vesicles reconstituted for active transport, *Proc. Natl. Acad. Sci. (USA)* **71**:1461.

Simoni, R. D., and Shallenberger, M. K., 1972, Coupling of energy to active transport of amino acids in *Escherichia coli*, *Proc. Natl. Acad. Sci. (USA)* **69**:2663.

Soderman, D. D., Germershausen, J., and Katzen, H. M., 1973, Affinity binding of intact fat cells and their ghosts to immobilized insulin, *Proc. Natl. Acad. Sci. (USA)* **70**:792.

Sprott, G. D., and MacLeod, R. A., 1972, Na$^+$-dependent amino acid transport in isolated membrane vesicles of a marine pseudomonad energized by electron donors, *Biochem. Biophys. Res. Commun.* **47**:838.

Sprott, G. D., and MacLeod, R. A., 1974, Nature of the specificity of alcohol coupling to L-alanine transport into isolated membrane vesicles of a marine pseudomonad, *J. Bacteriol.* **117**:1043.

Steck, T. L., and Wallach, D. F. H., 1970, The isolation of plasma membranes, in: *Methods in Cancer Research*, Vol. 5 (H. Busch, ed.), p. 93, Academic Press, New York.

Taube, R. A., and Berlin, R., 1972, Membrane transport of nucleosides in rabbit polymorphonuclear leukocytes, *Biochim. Biophys. Acta* **255**:6.

Van Thienen, G., and Postma, P. W., 1973, Coupling between energy conservation and active transport of serine in *Escherichia coli, Biochim. Biophys. Acta* **323**:429.

Venuta, S., and Rubin, H., 1973, Sugar transport in normal and rous sarcoma virus-transformed chick-embryo fibroblasts, *Proc. Natl. Acad. Sci. (USA)* **70**:653.

Wallach, D. F. H., 1967, Isolation of plasma membranes of animal cells, in: *The Specificity of Cell Surfaces* (B. D. Davis and L. Warren, eds.), p. 129, Prentice-Hall, Englewood Cliffs, N.J.

Wallach, D. F. H., and Kamat, V. B., 1966a, The contribution of sialic acid to the surface charge of fragments of plasma membrane and endoplasmic reticulum, *J. Cell Biol.* **30**:660.

Wallach, D. F. H., and Kamat, V. B., 1966b, Preparation of plasma membrane fragments from mouse ascites tumor cells, in: *Methods in Enzymology*, Vol. 8 (V. Ginsburg and E. Neufeld, eds.), p. 164, Academic Press, New York.

Wallach, D. F. H., and Lin, P. S., 1973, A critical evaluation of plasma membrane fractionation, *Biochim. Biophys. Acta* **300**:211.

Wallach, D. F. H., and Zahler, P. H., 1966, Protein conformations in cellular membrane, *Proc. Natl. Acad. Sci. (USA)* **56**:1552.

Wallach, D. F. H., Kamat, V. B., and Gail, M. H., 1966, Physicochemical differences between fragments of plasma membrane and endoplasmic reticulum, *J. Cell Biol.* **30**:601.

Warren, L., Glick, M. C., and Nass, M. K., 1966, Membranes of animal cells. I. Methods of isolation of the surface membrane, *J. Cell. Physiol.* **68**:269.

Weaver, R. A., and Boyle, W., 1969, Purification of plasma-membranes of rat liver: Application of zonal centrifugation isolation of cell membranes, *Biochim. Biophys. Acta* **173**:377.

Weiner, J. H., 1974, The localization of glycerol-3-phosphate dehydrogenase in *Escherichia coli, J. Membr. Biol.* **15**:1.

Weissbach, H., Redfield, B., and Kaback, H. R., 1969, Nucleotide binding by *Escherichia coli* membranes and solubilized membrane proteins, *Arch. Biochem. Biophys.* **135**:66.

Wheeler, K. P., and Christensen, H. N., 1967, Role of Na$^+$ in the transport of amino acids in rabbit red cells, *J. Biol. Chem.* **242**:1450.

Wu, H., Meezan, E., Black, P. A., and Robbins, P. W., 1969, Comparative studies on the carbohydrate-containing membrane components of normal and virus-transformed mouse fibroblasts. I. Glucosamine labeling patterns in 3T3, spontaneously transformed 3T3, and SV40-transformed 3T3 cells, *Biochemistry* **8**:2509.

Chapter 4

Electrophysiological and Optical Methods for Studying the Excitability of the Nerve Membrane

ICHIJI TASAKI and KENNETH SISCO

Laboratory of Neurobiology
National Institute of Mental Health
Bethesda, Maryland

1. INTRODUCTION

This chapter is devoted mainly to methods of preparing various types of excitable tissue for electrophysiological and optical investigations. Attempts are also made to explain how several different types of electrophysiological observations can be made using such tissue. In writing this chapter, the authors have assumed that most of the readers have never witnessed nerve fiber dissection of frog or squid. The authors had two objectives in preparing this chapter: (1) to give some idea of the technical problems which may be encountered by those who plan to work in this field of biological science and (2) to provide insight to investigators who are presently facing difficulties of some kind in their neurobiologically oriented research. The illustrations are designed to give readers a clear and correct image of the procedures involved.

Section 2 of this chapter concerns the isolation of the *Nitella* cell, single frog nerve fibers, and squid giant axons. A detailed description of internal perfusion of the squid giant axon follows in Section 3. Finally, principles and techniques are described in Sections 4–7 concerning measurement of optical properties of the squid axon, lobster nerve, and crab nerve during an action potential. These include fluorescence intensity changes deriving from dye molecules located in or near excitable membranes of squid axons or crab nerves and changes in the birefringence and light-scattering properties of nerves.

2. ISOLATION OF SINGLE EXCITABLE CELLS AND NERVE FIBERS

2.1. *Nitella*, an Excitable Plant Cell

The simplest means of observing propagation of action potentials of the all-or-none character is to use a single cell of *Nitella*. These cells are very large (often 1 mm in diameter and several centimeters long); hence isolation of a single cell does not require any surgical skill or training. The observed action potentials are large in amplitude and very long in duration; therefore, there is no requirement of an elaborate electrical device to demonstrate excitability. The membrane of a *Nitella* cell possesses essentially all of the properties of the nerve fiber membrane.

The discovery of formation *de novo* of an excitable membrane on the surface of a droplet of *Nitella* protoplasm (Inoue *et al.*, 1971) had a great impact on our understanding of the process of nerve excitation. This droplet offers an excellent possibility of studying properties of the membrane without the complication imposed by the presence of adhering satellite cells (Schwann cells, etc.) which exist outside nerve cells and fibers.

Nitella can be collected in freshwater ponds or streams. In the United States, it can be purchased from several commercial firms. There are many species of *Nitella*; however, it is difficult to identify the species of a particular specimen without the advice of a taxonomist. *Nitella flexilis* is best known because of the famous observations made by Osterhout and his associates (Osterhout and Hills, 1938). However, other species are equally usable for demonstration of excitability in single cells

Nitella can be kept almost indefinitely in the laboratory. The following artificial pond water is recommended: 0.025 mM KCl, 0.05 mM NaCl, 0.05 mM NaH_2PO_4, 0.2 mM $Ca(NO_3)_2$, 0.1 mM $MgSO_4$ (Inoue *et al.*, 1973). The portion of *Nitella* between two successive branching points (see Fig. 1A) is a single excitable cell known as an internode. Removal of the neighboring internodes by sectioning with a pair of scissors (without

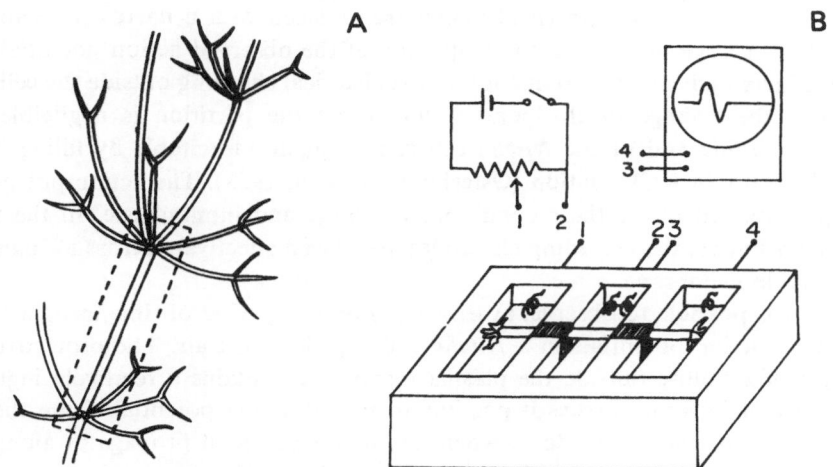

Fig. 1. (A) A portion of *Nitella* and (B) a simple experimental setup used to demonstrate action potentials developed by a single *Nitella* cell. The numerals indicate the connection of the silver–silver chloride electrodes to the stimulating and recording devices.

pinching or crushing the cell to be studied) completes the required surgery. Observation under a microscope reveals the existence of closely packed chloroplasts underneath the cellulose wall outside. Streaming of protoplasm (i.e., cyclosis) can also be seen below the layer of chloroplasts.

A lucite chamber provided with two partitions (see Fig. 1B) may be used to demonstrate propagation of an action potential along an internodal cell. Each partition has a narrow slit (about 2 mm wide) in the middle. The chamber is filled with artificial pond water deep enough to submerge the internodal cell. After the cell has been placed in the chamber, the partitions are sealed with a small amount of vaseline. Coils of silver wire— preferably chloridized (Robinson and Stokes, 1959) by passing an electrical current of the order of 10 C/cm² beforehand—are immersed in the pools of pond water in the chamber. A stimulating device—a battery in conjunction with a potentiometer and a switch—is connected to a pair of electrodes across one partition; a recording device (an oscilloscope) is connected to the electrodes across the other partition. An action potential several millivolts in amplitude can be evoked by manually closing the switch for a brief period of time. A commercially available stimulator is more convenient for this purpose than a manually operated battery circuit. Once an action potential is evoked, the cell becomes refractory to the second stimulating pulse. Full-sized action potentials can be evoked only at intervals longer than 20–30 s at 15°C.

If the partitions described above are replaced with a narrow air-gap, there is a large increase in the amplitude of the observed action potential. This increase is due to a reduction of the electrical shunting outside the cell. When the leakage of the pond water across the partition is negligible, the recording end of the *Nitella* cell can be made inexcitable by filling it with an 0.1 M KCl solution (Osterhout and Hill, 1938). The action potentials observed under these conditions are large and monophasic. In these circumstances, the recording electrodes are placed effectively across a single excitable membrane.

It is possible to observe, at least for a limited period of time, propagation of action potentials in a *Nitella* cell kept in moist air. The protective layer of cellulose outside the plasma membrane contains a relatively high density of ions and makes it possible to record action potentials from the surface. However, the rate at which an action potential propagates along the cell is very low under these conditions. This reduction in the rate of propagation is due to a decrease in the intensity of the electric current (known as local current) that exists between the resting and excited parts of *Nitella* plasma membrane. Continuous evaporation of water from the cell surface gradually decreases the high hydrostatic pressure inside the cell.

The interior of a *Nitella* cell is filled with vacuolar sap. The protoplasm constitutes only a thin layer between the cellulose wall and the vacuole. There is a continuous cyclic movement (cyclosis) in the layer of protoplasm facing the vacuolar sap. When a *Nitella* cell is cut across in a potassium-rich solution, the portion of the protoplasm involved in cyclosis slowly flows out of the cell. The surface layer of a protoplasmic droplet produced by this process is electrically inexcitable. When, however, such a droplet is transferred into a proper saline solution, the surface membrane of the droplet becomes electrically excitable within a period of 30–60 min. This excitability can be demonstrated by inserting a pair of hyperfine glass-pipette microelectrodes (one for stimulation and the other for recording) into a droplet of 100–300 μm diameter. The composition of the initial (potassium-rich) solution is as follows: 70 mM KNO_3, 50 mM $NaNO_3$, 5 mM $CaCl_2$. The composition of the final saline solution in which a new, electrically excitable membrane is formed on the surface of the naked protoplasm is as follows: 0.5 mM KNO_3, 0.5 mM NaCl, 1 mM $Ca(NO_3)_2$, 2 mM $Mg(NO_3)_2$.

It is now known that during the course of development of an excitable membrane there is an enormous change in the refractive index of the superficial layer and in the hydrostatic pressure inside the droplet (Inoue *et al.*, 1973).

2.2. Isolation of a Single Vertebrate Nerve Fiber

Vertebrate nerves (e.g., a sciatic nerve) contain a large number of nerve fibers which are the units capable of carrying action potentials. Compared to a *Nitella* cell or a squid giant axon (see below), vertebrate nerve fibers are very small, usually within the range of 2–15 μm in diameter. Because of their smallness, the technique of isolating a single vertebrate nerve fiber is tedious and a considerable amount of practice is required before a beginner becomes proficient enough to obtain reliable results. When this technique was developed in Japan before World War II (Kato, 1934; Tasaki, 1939, 1953), some European physiologists thought that it would be impossible for them to learn it. However, following Stämpfli's success (see Passo and Stämpfli, 1969), many Western investigators have mastered the technique. Originally the technique was developed to isolate a single frog or toad nerve fiber (see below). However, with slight modifications, it is possible to apply this method to mammalian nerve fibers. Historically, it was from rabbit phrenic nerves that single fiber action potentials were first observed (Adrian and Bronk, 1928).

The operation of isolating a frog nerve fiber is carried out in a shallow pool of Ringer's solution kept on a glass plate (approximately 10 by 5 cm). A dissecting microscope provided with an attachment for dark-field illumination is required. A pair of small scissors, small forceps (tweezers), and dissecting needles are used for the operation. In the following, isolation of a single motor nerve fiber innervating a frog gastrocnemius muscle is described. Gastrocnemius (a large muscle on the back side of the leg) is innervated as a rule by three small bundles of nerve fibers deriving from the tibial nerve (see Fig. 2A). When the connective tissue near the medial side of the muscle is carefully removed, the innervating small nerve branches can easily be seen entering the muscle. Then two of these branches are cut across, leaving only one of them. At this stage, electrical stimulation of the tibial nerve produces a vigorous twitch in the muscle. Within the tibial nerve, both the fibers entering the muscle and those running toward the Achilles tendon are surrounded by a single common layer of connective tissue sheath. When this common sheath is removed, it is possible to separate the motor nerve fibers innervating the muscle from one another by manipulating the fibers running toward the tendon.

The common connective tissue sheath can be split by using either dissecting needles (Fig. 2B,C) or a pair of small scissors. During this procedure, it is important to keep the nerve and the muscle in their natural (not twisted) relative positions. Then by pulling the ends of the cut nerve

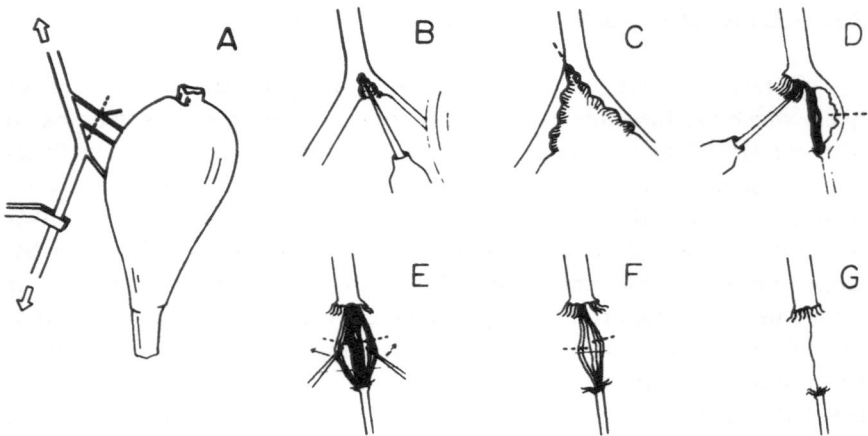

Fig. 2. Procedure of isolating a singe nerve fiber from a sciatic-gastrocnemius preparation.

fiber on one side and the surface layer of the sheath on the other side, the bundle of the nerve fibers entering into the muscle is separated from the sheath (D). Next, by inserting the tips of two (slightly blunt) needles directly into the desheathed nerve branch, intact nerve fibers in the middle are separated from one another (E). Finally, all but one (large, healthy looking) fiber are cut across with a needle (F). When a motor nerve fiber is successfully isolated, stimulation of the tibial nerve evokes a visible twitch in the gastrocnemius muscle.

The dissecting needles used in this operation need not have sharp cutting edges. Pressing the surfaces of the glass plate with the sharp tip of a needle is a very effective means of perforating a strong layer of connective tissue. Periodic pressing combined with a gradual withdrawal of motion (intending to draw a fine dotted line on the glass surface with the needle tip) readily cuts the sheath around the nerve trunk.

During this operation, there is continuous evaporation of water on the glass plate. It is therefore important to replenish the saline solution on the glass plate frequently. However, it is difficult to carry out this type of operation in a deep layer of Ringer's solution because nerve fibers tend to move around uncontrollably and come together under such circumstances.

In order to record action potentials from these fibers, it is necessary to transfer the fibers to a special recording setup. During such a procedure, the isolated portion of the nerve fiber should not be stretched beyond its natural length. Furthermore, the isolated portion should not be lifted above the surface of the Ringer's solution. (Note that the electrostatic field in the room frequently generates a current strong enough to damage the fiber.) The preparation can readily be transferred from one plate to another by keeping the fiber floating in a layer of Ringer's solution.

Several different methods have been developed for recording action potentials from a single myelinated nerve fiber (see Passo and Stämpfli, 1969; Tasaki and Frank, 1955; Frankenhaeuser, 1957). The reader interested in these special methods is referred to the original articles. For investigators who are interested only in counting the number of action potentials (and not their size or shape), the classical method developed by Adrian and Bronk (1928) and by Hartline (1938) is recommended.

2.3. Crab or Lobster Nerve Fibers

Nerve trunks and fibers of various species of crabs have been used for a variety of purposes in neurophysiology. Dissection of these nerve trunks is relatively easy and isolation of single nerve fibers of these animals requires little or no practice.

A claw of a spider crab (*Libinia emarginata*) may be removed by crushing its attachment to the body with a pair of strong scissors or a rongeur (Fig. 3A). To expose the nerve, removal of the exoskeleton on the medial side (see Fig. 3B) is recommended. The claw nerve is located beneath a flat tendon near the middle joint. By inserting a blade of scissors under the tendon, the end of the muscle covering the nerve can be severed. Then the entire muscle can be pulled away, exposing the major portion of the nerve. At this stage, immersion of the claw in seawater is desirable. The nerve can be removed after tying the two ends with thread.

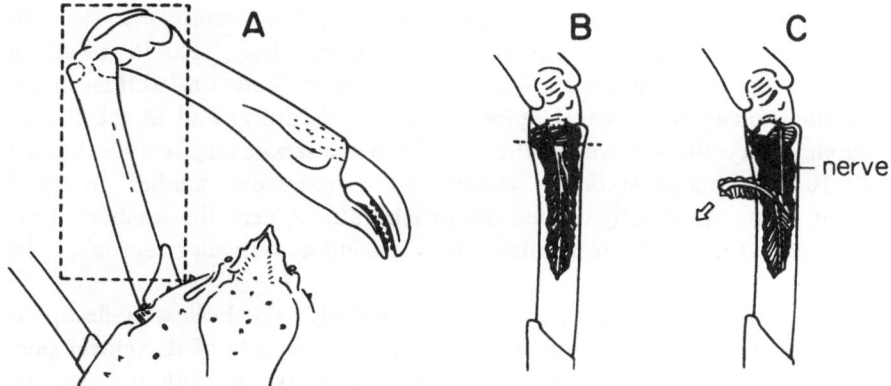

Fig. 3. An approach to a nerve trunk innervating the claw of a crab.

When a freshwater crab or crayfish is used instead of a *Libinia*, a physiological saline solution which has a lower ionic strength must be used in place of artificial seawater. The most commonly used one is a van Harreveld solution (see Cavanaugh, 1956) having the following composition: 205.3 mM NaCl, 5.37 mM KCl, 13.55 mM $CaCl_2$, 2.61 mM $MgCl_2$, 2.38 mM $NaHCO_3$.

The exoskeleton of a lobster (*Homarus americanus*) is relatively soft; hence it can be cut easily with a pair of scissors. The nerve innervating a walking leg is located roughly at the center of the leg; it can readily be exposed after removing one-half of the exoskeleton. To remove the potassium salt that comes out of the damaged muscles, repeated rinsing of the exposed nerve with running seawater is recommended. The nerves innervating two pairs of legs in front are larger and longer than those innervating other legs.

Leg nerve fibers of a lobster or a crab can be isolated by a very simple method invented by the Japanese physiologist Furusawa (1929). This method consists of pulling the nerve out of the meropodite after cutting the ligament around the joint. The nerve fibers isolated by this method are known to be in excellent physiological condition.

To observe the action potential of these fibers extracellularly, a lucite chamber similar to that illustrated in Fig. 1B may be used. However, since the duration of these action potentials is far shorter than that of a *Nitella* cell, brief stimuli, preferably shorter than about 0.5 ms, are needed for stimulation.

2.4. Squid Giant Axons

Squid which possess giant axons (*Loligo, Doryteuthis*, etc.) are encountered in various parts of the Atlantic Ocean (e.g., near Plymouth in England and Woods Hole, Mass., in the United States) as well as in the Mediterranean Sea (near Naples and Genoa in Italy) and in the Sea of Japan. Since the survival of these squid in the laboratory is quite limited (3–10 days under favorable conditions), experimental studies on squid giant axons are usually carried out in laboratories near the seashore. Furthermore, the squid are available only during a particular season of the year.

Dissection of a squid giant axon is relatively easy. Following decapitation, the mantle of the animal is opened in the middle of its ventral side. Caution has to be taken to avoid cutting the ink sac with the scissors. Then the digestive and reproductive organs are removed without damaging

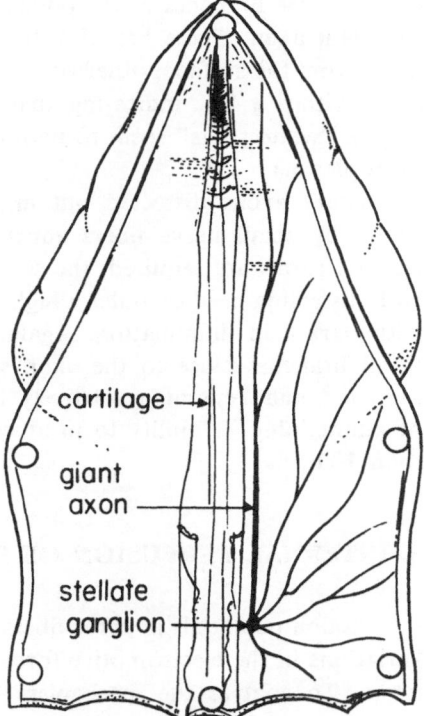

Fig. 4. Position of the giant axon on the inner side of a squid mantle.

cartilage →

giant
axon

stellate
ganglion

the giant axons, which are located near the middorsal wall of the mantle (Fig. 4). This operation is carried out by moving the tip of the scissors along the surface of the thin cartilage on the dorsal wall of the body cavity. If the specimen is fresh, the mantle is transparent and the giant axon running along the edge of the cartilage can easily be recognized by the unaided eye. Under a low-power dissecting microscope used in conjunction with dark-field illumination, the giant axons are separated from the surrounding tissue and small fibers with a pair of small scissors. It is recommended that the entire operation be carried out with the mantle of the animal submerged under running seawater.

Detailed examination of a giant axon under a dissecting microscope during the course of dissection reveals the existence of several small branches of the axon entering the underlying muscle (see, e.g., Arnold et al., 1974). When one of these branches is cut during dissection, a small localized twitch is observed in the muscle. Some of these branches are relatively large, sometimes reaching 50 μm in diameter. It is important to avoid cutting these branches at the points close to their origin on the surface of the giant axon. These branches have to be cut at points at least 0.4 mm away from the surface; otherwise, small "white spots" develop in the giant axon within $\frac{1}{2}$–1 h, indicating that the giant axon is severely damaged. (These "white spots" seem to be produced by the entry of $CaCl_2$ into the protoplasm.)

Giant axons dissected out in this manner are surrounded at least partly by small nerve fibers adhering to the surface. When extensively cleaned axons are required, the surrounding small fibers may be removed with dissecting needles under a high-power dissecting microscope combined with dark-field illumination. Again, caution has to be taken not to cut small branches close to the axon surface. A giant axon, when properly prepared and kept at a relatively low temperature (5–10°C), is capable of maintaining its ability to produce full-sized action potentials for more than 10 h.

3. INTERNAL PERFUSION OF SQUID GIANT AXONS

Action potentials in nerve fibers are produced as the result of transient variations in the electromotive force across the surface membrane of these fibers. To analyze the mechanism of development of this electromotive force from a physicochemical point of view, it is absolutely necessary to be able to alter the chemical compositions of the fluid media on both sides of the membrane. It is very easy to alter the fluid medium outside. However, until fairly recently it was not known how to manipulate the chemical composition of the fluid medium inside at will. In 1961, methods of internal perfusion (Oikawa *et al.*, 1961; Baker *et al.*, 1961) were devised which enabled us to observe action potentials that developed across the axonal membrane separating two electrolyte solutions of known chemical compositions. Up to the present time, these methods have been successfully applied only to squid giant axons. There seems no reason why these methods cannot be applied to giant axons of *Myxicola infundibulum* (see Binstock and Goldman, 1969). Attempts to perfuse frog muscle fibers (Davies, 1961) and crayfish axons internally (Shrager *et al.*, 1969) met with only limited success.

There are two different methods of internal perfusion, one developed at Plymouth, England, and the other at Woods Hole, Mass. The British method consists of squeezing out the protoplasm from a cut end of the axon with a small rubber roller and later filling the axon interior with an artificial solution. In the American method, which will be described below in some detail, flow of the internal solution is maintained by using two glass cannulas inserted into the axon at the end. Some investigators (e.g., Fishman, 1970) have used only one cannula instead of two (without squeezing out the protoplasm). With this single-cannula method, however, it seems difficult to maintain a continuous flow of the internal solution, or to replace one solution with another swiftly.

3.1. Perfusion Chamber

In order to perfuse a squid giant axon internally with two glass cannulas, it is necessary to maintain the axon in a horizontal position. Furthermore, the site of incision of the axon membrane, where the cannulas are introduced into the axon, should not be immersed in the external fluid medium, because the axonal membrane is readily destroyed by diffusion of seawater into the axon. The chamber shown in Fig. 5 was devised to satisfy these requirements (see Tasaki, 1968).

This chamber is made of transparent lucite. The main part is 25–35 mm wide and approximately 50 mm long. There are two side pieces which serve to raise the axon slightly (about 0.5 mm) above the lucite surface of the chamber. There are two pairs of silver (or platinum) electrodes near the edges of the main chamber, one for delivery of brief electrical shocks to the axon and the other for recording action potentials extracellularly.

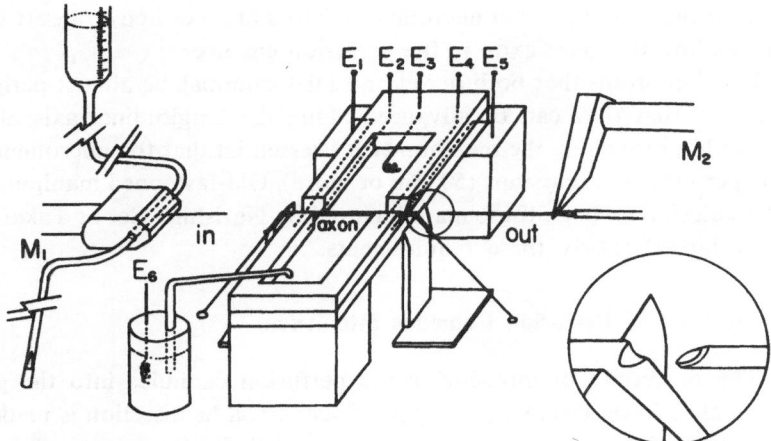

Fig. 5. Experimental setup used for internal perfusion of a squid giant axon. The inlet and the outlet cannulas are represented by "in" and "out," respectively; M_1 and M_2 are fixed to micromanipulators; Es represent electrodes.

The side pieces are fixed at a distance of 2–3 mm away from the edges of the main chamber. Both the surface of the side pieces and the edges of the main chamber are made water repellent by coating with melted paraffin. A shallow pool of seawater (2–3 mm deep) is made in the chamber.

3.2. Inlet and Outlet Cannulas

When a cannula is introduced into a squid giant axon, the hydrostatic pressure inside the axon tends to rise. Because of the injurious effect of this pressure, it is not possible to push glass tubing with a diameter exceeding about 100 μm into an axon without suppressing excitability. If, however, a large portion of the protoplasm is removed beforehand, a large glass cannula can be introduced without bringing about any harmful effect. Furthermore, a large channel made in the protoplasm by suction profoundly reduces the probability of obstruction of the flow of the perfusion fluid. For these reasons, the perfusion system shown in Fig. 5 was devised.

Both the inlet and outlet cannulas are made of glass. The outside diameter of the inlet cannula is approximately 90 μm and that of the outlet cannula is roughly 300 μm. The tip of the outlet cannula is beveled and rounded with a rotating whetstone, as shown in Fig. 5. The tip of the inlet cannula is simply rounded. It is held by a plastic rod and is connected to a reservoir of the perfusion solution in such a manner that there is very little dead space. The outlet cannula is fixed to a separate plastic rod as shown in the diagram. Two micromanipulators are required to insert these cannulas into the giant axon in the perfusion chamber.

It is important that both outlet and inlet cannulas be almost perfectly aligned so that they can be advanced along the longitudinal axis of the axon without touching the membrane. It is essential that the micromanipulators permit long excursions (50 mm or more). Old-fashioned manipulators of the Peterfi type (e.g., those manufactured by Narishige Co. or Takahashi Co. in Japan) satisfy these requirements.

3.3. Insertion of Perfusion Cannulas into Axon

The procedure of introducing the perfusion cannulas into the giant axon is as follows. Using a small pair of scissors, a hemisection is made on the portion of the axon membrane above one of the side pieces. Then the tip of the outlet cannula is inserted into the opening in the membrane. Next the cannula is slowly advanced along the longitudinal axis of the axon. As the cannula is advanced, the axoplasm is continually sucked into it, so that there is no visible increase in the diameter of the axon. This is done by connecting polyethylene tubing to the bottom end of the cannula and applying gentle suction by mouth. When the cannula has a uniform diameter and is properly beveled and rounded at the tip (see Fig. 5, inset), it is amazingly easy to introduce a 20- to 30-mm-long portion of the cannula into the axon without bringing about any noticeable decrease in the rate of nervous conduction.

The next step in the procedure is to introduce the inlet cannula into the axon at the other end. When the alignment of the two cannulas is sufficiently good, it is possible to insert the tip of the inlet cannula into the lumen of the outlet cannula. This step is relatively easy if the orifice of the outlet cannula is facing upward. Then, with a high hydrostatic pressure applied to the fluid in the inlet cannula, a flow of the perfusion fluid is initiated by gentle suction applied to the outlet cannula. (At this stage, there is no direct contact between the protoplasm of the axon and the flowing perfusion fluid.) Finally, with the flow maintained between the two cannulas, their tips are separated, bringing the axon interior in direct contact with the flowing perfusion fluid. The action potential of the axon under internal perfusion can be monitored by inserting a small glass pipette electrode (roughly 70 μm in diameter) into the axon interior through the outlet cannula.

3.4. Chemical Composition of Internal Perfusion Fluid

In the early stage of development of the technique of internal perfusion, an isotonic solution of potassium chloride (0.5 M KCl) was used as the standard perfusion fluid. It was found that with this solution flowing inside an axon the action potential disappeared regularly 25–40 min after the onset of internal perfusion (Oikawa et al., 1961). Later, the significance of this finding was clarified. It was shown that the time between the onset and the disappearance of action potentials is determined almost exclusively (1) by the lyotropic numbers of the internal anions and cations and (2) by the salt concentration of the internal solution (Tasaki et al., 1965). The following sequence of anions arranged according to their lyotropic number agrees perfectly with the order determined on the basis of "favorability" as internal anions: fluoride $\geq$ phosphate $>$ sulfate $>$ chloride $>$ bromide $>$ iodide $>$ thiocyanate, where fluoride and phosphate are the most favorable anions.

Dilution of the internal perfusion fluid with 1.1 M sucrose or 12%

(by volume) always has a favorable effect on the axon excitability. For example, at the moment when nervous conduction is blocked as the result of internal perfusion with an isotonic KCl solution, dilution of the internal perfusion fluid with the nonelectrolyte solution by a factor of 2–4 is known to restore excitability immediately.

Among common univalent cations, cesium is most favorable; sodium is not very favorable. Potassium and rubidium are intermediate between Cs and Na. However, nervous conduction can be maintained for a long period of time even with Na ions internally if combined with fluoride. Multivalent cations, Ca ions in particular, are extremely poisonous when applied intracellularly.

As a standard internal perfusion fluid for observing "normal" action potentials, use is recommended of a mixture of 4 parts of 0.55 M KF solution and 1 part of 12% (by volume) glycerol solution having its pH adjusted to 7.2–7.4 with a small amount of potassium of phosphate buffer solution (0.04 M with respect to K). (In preparing internal perfusion solutions, the use of double glass-distilled water is recommended.) As an external medium, either natural or artificial seawater can be used. The artificial seawater used in Woods Hole (Cavanaugh, 1956) has the following composition: 423 mM NaCl, 9 mM KCl, 9.27 mM $CaCl_2$, 22.9 mM $MgCl_2$, 25.5 mM $MgSO_4$, and 2.15 mM $NaHCO_3$ or 7.3 mM tris chloride (titrated to pH 8.0 with HCl). A mixture of 1 part of 0.4 M $CaCl_2$ solution and 5 parts of 0.6 M NaCl solution having its pH adjusted to 8 with a tris buffer may also be used (about 10 ml of 0.73 M tris chloride, pH 8, is needed for 1 liter of solution).

3.5. Some Experimental Findings Obtained by the Technique

When the technique of intracellular perfusion was invented, the first experiment carried out using it was determination of the dependence of the resting membrane potential on the internal potassium ion concentration. It was postulated by Bernstein (1902) that the resting potential, E_r, is determined by the Nernst equation applied to the potassium ion concentration across the membrane:

$$E_r \text{ (in mV)} = 58 \log \frac{[K]_o}{[K]_i}$$

where $[K]_o$ and $[K]_i$ are the external and internal potassium ion concentrations, respectively. If this is true, a 10-fold reduction of $[K]_i$ is expected to lower the membrane potential by 58 mV. The change observed directly by the method of internal perfusion was about 10 mV (Tasaki *et al.*, 1962; Baker *et al.*, 1962).

The experimental result just described is consistent with the view held by several investigators (Stämpfli, 1959; Tasaki and Singer, 1966; Ling, 1960) that the resting membrane potential is not determined by the Nernst equation cited above. The effect of varying $[K]_o$ is also known to be inconsistent with the behavior expected from the simple Nernst equation. In the range of $[K]_o$ which does not suppress the ability of the axon to produce action potentials, a large change in $[K]_o$ does not affect the resting membrane potentials.

In the experiments described above, the internal salt concentration was varied by mixing a potassium salt solution with a nonelectrolyte (sucrose or glycerol) solution. Therefore, there is in the axon interior a large variation in the ionic strength which may influence the macromolecular state of the axon membrane. The effects of varying the ionic strength can be avoided by changing the internal potassium salt concentration by mixing it with a sodium salt solution. In such experiments, there are complications arising from the tendency of the internally applied Na ions to produce irreversible deterioration of excitability. To avoid such an "unfavorable" internal effect of Na ions, the sum of $[Na]_i$ and $[K]_i$ was maintained at a level of about 0.1 M (sometimes 0.05 or 0.2 M). It was found by this procedure (Tasaki, 1968) that the resting membrane potential is totally insensitive to a wide variation of $[K]_i$.

All these experimental results are quite consistent with the notion that the axon membrane has fixed negative charges of a low density on the inner surface of the axon membrane. In accordance with Teorell's (1953) theory of charged membranes, the potential difference across the axon membrane at rest consists of (a) a phase-boundary potential at the outer surface, (b) a diffusion potential within the membrane, and (c) a phase-boundary potential at the inner surface. There is no doubt that the contribution of the third component is relatively small.

The effect of varying $[Na]_i$ on the action potential was examined by the use of the technique of internal perfusion. One of the most interesting findings along this line is the demonstration of all-or-none action potentials in axons internally perfused with a dilute NaF solution and immersed in a medium containing $CaCl_2$. Figure 6A shows an example of the records demonstrating propagation of such an action potential. In this case, the external $CaCl_2$ concentration was close to the total divalent cation concentration in normal seawater and the internal Na ion concentration was not far from that in the normal axoplasm. Therefore, the cation composition in

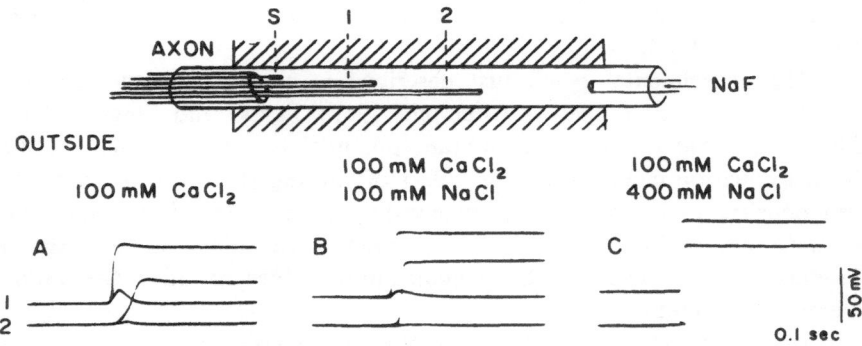

Fig. 6. Simultaneous recording of action potentials of a squid giant axon with two internal glass-pipette electrodes (1 and 2). The axon was internally perfused with a dilute NaF solution. The electrolyte compositions of the external media used are indicated. Electric stimuli were delivered through an internal metal electrode (S). From Inoue *et al.* (1974).

this system was very similar to that of a normal membrane except that both the external sodium salt and the internal potassium salt were replaced with nonelectrolyte. Therefore, this demonstration may be regarded as a piece of evidence that neither the external Na nor the internal K ions are essential for propagation of all-or-none action potentials. Of course, these cations do exert strong influence on the amplitude and the duration of the action potential. Insofar as the present authors are aware, the action potential of a squid giant axon cannot be maintained in the absence of divalent cation in the external medium. In this respect, external divalent cations (Ca ions in particular) are essential for production of action potentials. Figure 6B,C shows the effect of NaCl added to the external medium under these conditions. As the external Na ion concentration approached the level in normal seawater, the rate of propagation was seen to approach the normal propagation rate in a squid giant axon.

A different type of application of the internal perfusion technique is to utilize it to investigate the nature of membrane macromolecules. Studies along this line are still in progress. However, it seems worthwhile to describe some salient facts discovered by the use of this technique.

The nature of the membrane macromolecules can be analyzed by using specific biochemical dissecting tools, namely, enzymes. It was found that the following proteinases bring about irreversible loss of excitability when applied internally at the level of about 0.1 mg/ml in a potassium phosphate solution: (1) trypsin, (2) chymotrypsin, (3) carboxypeptidase, (4) leucine aminopeptidase, (5) papain, and (6) ficin. It is extremely interesting that these proteinases have practically no effect on excitability when applied to a squid giant axon externally (see Tasaki, 1968).

Phospholipase A suppresses excitability when applied either externally or internally. Phospholipase C suppresses the ability of the axon to produce action potentials when applied internally. The effect of externally administered phospholipase C is dubious.

These experimental findings strongly suggest that the axonal membrane is made of labile lipoprotein complexes whose conformations are extremely sensitive to the ratio of the divalent–univalent cation concentration ratio in the medium. At present, the most reasonable physicochemical explanation of the process of nerve excitation is to attribute it to a rapid, reversible conformational change of the membrane lipoprotein complexes.

4. PHYSICAL PRINCIPLES INVOLVED IN OPTICAL MEASUREMENTS

Action potentials may be regarded as an electrical manifestation of a rapid physicochemical change taking place in the membrane macromolecules. In order to elucidate the physicochemical basis of the process of action potential production in the nerve, it is highly desirable to study various properties of the nerve membrane by optical (i.e., nonelectrical) means. Until recently, however, nobody had succeeded in demonstrating changes in optical properties of the nerve membrane during nerve excitation.

In 1968, optical signs of a rapid, reversible conformational change in the membrane macromolecules were discovered by using the birefringence technique, the method of light scattering, and the method of extrinsic fluorescence (Cohen *et al.*, 1968; Tasaki *et al.*, 1968). This success opened up an entirely new field of investigation. In this and the following sections, we are concerned with these optical methods applied to various nervous tissue.

There are physical problems that are common to all the optical measurements on the nerve. In these measurements, we are dealing with small changes in strong background light. The ratio of the changes to the background intensity is usually between $1:10^4$ and $1:10^5$. Furthermore, since action potentials are ordinarily between 1 and 10 ms in duration, a relatively high time resolution is required in these measurements. This situation creates special problems which are not commonly encountered in ordinary physical measurements.

In measurements of fluorescence signals, it is difficult to use an incandescent lamp as the light source because the ultraviolet content in the emitted light is usually too weak. On the other hand, the light intensities emitted by tubes of the gas-discharge type (e.g., xenon or xenon-mercury types) are quite unstable. This instability creates special problems in the optical recordings. In addition, one encounters a number of technical problems which are not described in most books on optics. These problems and several methods of overcoming the difficulties are discussed in the following section.

4.1. Random Noise

A serious problem in the detection of a small change in strong background light is the presence of irregular noises of the recording system which persist in the absence of a physiological signal. The presence of a signal can be recognized only when the signal amplitude is larger than the noise amplitude. There are three kinds of undesirable deflections (noises) in our recording system: (1) random (white) noise, (2) incidental machine noise, and (3) 60-cps noise and its higher harmonics deriving from the electrical power supplies of the optical recording systems. We shall first deal with random noise.

It is possible to estimate the number of photons entering into our recording system. Suppose the number happens to be 10^{13} photons per second and the quantum yield of the detector is 10%. Then the number of electrons emitted from the cathode of the photomultiplier (or the photodiode) is, on the average, 10^{12}/s. If the time resolution required for our recording is $\frac{1}{3}$ ms (e.g., when recorded with a computer in which there are 100 dots in 30 ms), the average number of photoelectrons emitted during this brief period of time is 3×10^8. Since the emission of electrons from the surface of the photocathode (or photodiode) is random, there is a fluctuation in number above and below this average value. The standard deviation for this fluctuation is equal to the square root of the average number, 1.7×10^4. In this case, therefore, unless the number of photoelectrons contributing to production of a physiological signal is larger than this number, the signal will be buried below the level of this random noise. Obviously this level of random noise cannot be reduced by lowering the temperature of the photodetector. This argument indicates that unless the background light level is stronger than about 10^{13} photons a physiological optical signal whose the amplitude is only 10^{-4} or less of the background light cannot be observed directly on the screen of a recording oscilloscope.

It is evident that random noise of this kind increases with the square root of the intensity of the incident light, while the signal amplitude increases linearly with the light intensity. It follows from this that the signal-to-noise ratio increases with the square root of the intensity. This is the reason why a strong light source is required to record optical signals from the nerve. A method of estimating the number of photons arriving at the photodetector is described in a book edited by Johnson and Haneda (1966). A rough determination can be made by estimating the number of the photo-electrons in the first stage of the photomultiplier and dividing this estimate by the well-known value 1.6×10^{-19} C/electron.

The quantum yield of a photomultiplier is known to be greatest when the applied voltage between the first dynode and the cathode is 200–300 V. Since each successive dynode increases the number of electrons by a factor of about 3, the output of an 11-stage photomultiplier will have $10^{12} \times 3^{11}$ electrons/s or 100 mA when 10^{13} photons are reaching the photomultiplier every second. This current level is too high for the photomultiplier anode. For this reason, the photomultipliers in our recording system (RCA C70109E or 4463) are used as a three- or four-stage multiplier by connecting all the last seven or eight pins to the anode.

4.2. Fluctuations in the Light Intensity from the Source

Commercially available optical instruments employed by biochemists usually have a very low time resolution. Therefore, fluctuations in light intensity from the source which occur within 1 s are totally unimportant for the manufacturers of those instruments. However, the optical signals which we are measuring from the nerve have a duration of between 1 and 10 ms. Therefore, a fluctuation in the light intensity with a periodicity of 120 cps (which arises from imperfection in the smoothening of the current from the power supply) can create a serious problem in optical measurements on nerves. There are also fluctuations in the light intensity due to a flickering of the arc when a xenon or xenon-mercury lamp is used as a light source.

The major portion of these fluctuations can be suppressed by connecting large storage batteries across the lamp terminals. A further suppression of fluctuation can be achieved by leading a small fraction of the light beam from the source to a reference photomultiplier and recording the difference in light intensity between the reference and detector channels. This differential method is effective in reducing the fluctuation by a factor of 100 or more when properly used. In a xenon or mercury lamp, it is important to divide the reference beam by a beam splitter (a quartz coverslip) placed in front of a nerve preparation.

4.3. Suppression of a 60-cps Sinusoidal Wave or Its Higher Harmonics in Averaging Computer Recordings

In recording small signals superposed on a random noise, a precaution has to be taken to eliminate 60-cps disturbances which may enter into the recording system at every stage between the photomultiplier and the computer. Since this 60-cps AC and its higher harmonics are the major source of trouble in our recording, it is important to consider the following points:

1. Electrical coupling between the recording system and the power line. This coupling can be reduced or eliminated by shielding the system with a metallic conductor (e.g., aluminum foil) which is connected to ground.
2. Magnetic fluxes affecting the recording system. This effect cannot be eliminated by insertion of an ordinary metal plate between the source and the recording system. It is easy to reduce this effect by increasing the distance between the source (e.g., a transformer in the power source) and the recording system. Furthermore, rotation of the power source or insertion of a ferromagnetic metal (mu-metal) sheet often is very effective in suppressing the disturbance.
3. Disturbances deriving from improper grounding of the power supplies. The disturbances of this type seem to be least understood by the investigators in the field. We are dealing here with an AC of the order of a small fraction of 1 mV which may be amplified and recorded by an averaging computer. Each power supply is a possible source of a 120-cps AC due to imperfect smoothening. This AC can easily be detected by connecting the differential inputs of an oscilloscope to different parts of the grounding wire of the instrument. At the level of sensitivity required for our recording, the grounding wire is not equipotential along its course from the computer to the photomultiplier. The best method for eliminating this disturbance is to ground every part of the instrument with a heavy wire at a single point. Double or multiple grounding must be avoided. It is extremely easy to have multiple grounding, because the power supplies, the amplifiers, the oscilloscopes, the computer, etc. are connected by shielded cables and the shielding can act as

an additional grounding wire. If these parts are mounted on a common metallic rack, the rack can act as an additional ground. Furthermore, a common power supply can introduce a ground loop into the system. To repeat, the failure to recognize the existence of multiple grounding appears to be the most common source of trouble.

4. Suppression of 60-cps AC or its harmonics in averaging signals. When the amplitude of the optical signal is smaller than the amplitude of the random noise, the use of an averaging computer is essential for the retrieval of the signal. After averaging 100 times, a signal which is one-tenth of the random noise becomes recognizable; following 10,000 averagings, such a signal becomes 10 times as large as that of the random noise. (Note that the signal amplitude increases directly with the number of additions, while the random noise rises with the square root of the number.) During the process of recording such small optical signals, 60-cps or 120-cps AC completely hidden below the level of the random noise can become recognizable if the frequency of repetition of the averaging computer is exactly equal to a subharmonic of 60 cps. Therefore, it is absolutely essential that the averaging computer not be operated at exactly 3, 6, 9, 12, 15, 18 cps, etc. If one repeats the averaging of the computer at a period given by

$$(n + \tfrac{1}{2}) \tfrac{1}{60} \text{ s}$$

where n is an integer, a disturbance which repeats at 60 cps is eliminated because a positive phase of a cycle of 60 cps is immediately followed by a negative cycle. Note, however, that a 120-cps disturbance is enhanced in this case. In this laboratory, repetition for squid experiments is usually at a rate close to 14.1 shocks per second, i.e., at an interval of approximately

$$(4 + \tfrac{1}{4}) \tfrac{1}{60} \text{ s}$$

This method appears to be effective in suppressing both 60- and 120-cps disturbances.

5. Sharp dependence of photomultiplier output on the applied voltage. A small fluctuation in this voltage produces a large fluctuation in the output. Note that this fluctuation is proportional to the light intensity. Additional smoothening of the power supply output is required in this case.

4.4. Mechanical Disturbances

The experimental setup for recording optical signals from the nerve is usually very sensitive to mechanical disturbances. Vibration of the floor or of the wall of the room, as well as sound waves, can strongly disturb the recording system. The major portion of these disturbances can be eliminated by placing the optical setup on a heavy plate supported by an air cushion, e.g., on a granite slab resting on inflated automobile inner tubes. Transmission of disturbances through the cables connecting the optical setup with the power supplies and recorders can be suppressed by taping the cables to the plate.

When the light is focused on the nerve, a small movement of the focusing lens relative to the nerve may create a large change in the intensity of light reaching the nerve. When the position of the lens is adjusted to yield a maximum of the light intensity, the optical setup is least sensitive to mechanical vibrations.

5. EXTRINSIC FLUORESCENCE SIGNALS

In recent years, the method of fluorescence labeling has been used to study conformational changes in various macromolecules (Stryer, 1968; Brand and Gohlke, 1972; Radda and Vanderkooi, 1972). This method has been applied to the nerve membrane in an effort to elucidate the process of action potential production (Tasaki *et al.*, 1972, 1973a,b). It is now established that when a nerve labeled with various fluorescent probes is electrically stimulated there is a transient change (either decrease or increase) in the intensity of the fluorescent light deriving from the probes. The fluorescence of these probes has been studied by externally labeling crab nerves or by internally labeling a squid giant axon. A large number of fluorescent probes have been shown to produce these small changes in fluorescence intensity in response to electrical stimulation of labeled nerves (Tasaki *et al.*, 1969b; Cohen *et al.*, 1974).

Most of our studies are now being directed toward understanding the molecular basis for these signals. We constructed a spectrofluorometer specifically suited for studies of the nerve membrane and carried out spectral analysis of a number of dyes (Tasaki *et al.*, 1973a). Such a special spectrofluorometer is needed because the time resolution of the commercially available machine is too low to be used for detection of the rapid changes in fluorescence occurring during nervous activity. We examined the polarization properties of fluorescence signals with a view toward elucidating the structure of the binding sites for probe molecules in the nerve (Tasaki *et al.*, 1974). A group of investigators led by L. B. Cohen is developing a technique for "visualizing" action potentials by the use of fluorescent dyes (Davila *et al.*, 1973).

5.1. Optical Setup for Measuring Fluorescence Signals

An optical setup for detecting fluorescence signals consists of a device to excite optically the probe molecules in the nerve and an arrangement to collect the emitted fluorescent light following stimulation. When the probe molecules incorporated in the nerve absorb visible light, a quartz-iodine lamp operated with storage batteries can be used as the light source. However, when all the absorption band of the probe molecule is located in the ultraviolet range, a more elaborate setup is required.

The diagram at the top of Fig. 7 shows the arrangement used to detect fluorescence signals from a crab nerve prestained with acridine orange (Tasaki *et al.*, 1969*a*). White light from a 150-W quartz-iodine lamp (Os-

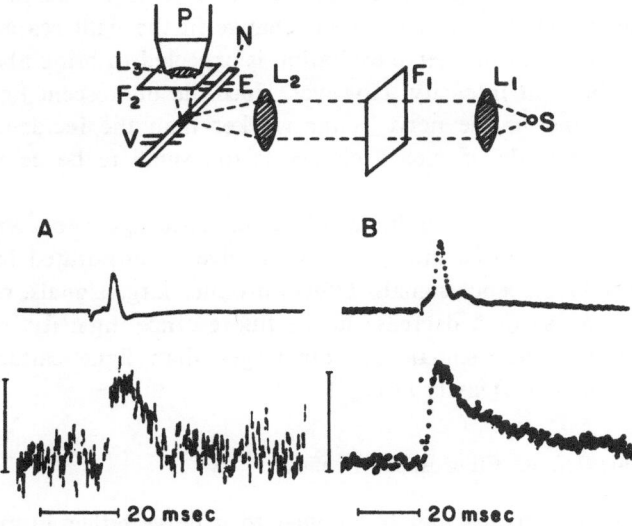

Fig. 7. Top: Optical setup used to measure changes in the intensity of extrinsic fluorescence during nerve excitation. Bottom: Records of fluorescence intensity changes (lower trace) associated with action potentials (upper trace) in a crab nerve stained with acridine orange. Record A was made from a single sweep of a dual-beam oscilloscope; record B was taken with an averaging computer. From Tasaki *et al.* (1969*a*).

tram), S, was converted into a quasi-monochromatic light beam focused on the nerve by means of lenses, L_1 and L_2, and an interference filter (465 nm), F_1. The fluorescent light from the nerve was measured with a photomultiplier, P, through a cutoff filter, F_2, which absorbs the exciting light but transmits the fluorescent light. Note that the wavelength of the fluorescent light is almost always longer than that of the exciting light. In response to electrical stimulation of the nerve, there was a small, transient increase in the light intensity. As can be seen Fig. 7, there was a strong noise superposed on the signal (observed directly on the screen of an oscilloscope). Signal averaging significantly reduced the noise level relative to the signal.

In this type of measurement, it is important to ascertain that the light transmitted through the primary filter (F_1) is almost completely blocked by the secondary filter (F_2). A simple test of this point is to move F_2 to a position between the F_1 and the nerve (N). By this procedure, the light intensity reaching the photomultiplier has to be reduced to a negligibly weak level if the two filters are properly chosen. It is also necessary to conduct a control experiment using an unstained nerve with both F_1 and F_2 in their proper positions.

Under the conditions of the experiment illustrated in Fig. 7, the fluorescent light deriving from the probe molecules in the nerve is scattered by the nerve membrane. A transient change in the light scattered by the nerve membrane during nerve excitation is expected to bring about a small change in the light intensity. However, since the fluorescent light deriving from the probes in the nerve is far weaker than the incident (exciting) light, the magnitude of such a change is too small to be detected under these conditions.

The signs and the magnitudes of fluorescence signals are known to be very different for probe molecules. Many dyes incorporated in the nerve produce no fluorescence signals. Others produce large signals, representing either an increase or a decrease in the fluorescence intensity. In some instances, fluorescence signals are far larger than light-scattering or birefringence signals (Davila *et al.*, 1973).

5.2. Preparation of Fluorescent Probes

Fluorescent probes can be applied to a nerve either intra- or extracellularly. When intracellular application of a probe is desired, it is dissolved in an isotonic potassium phosphate solution (pH 7.3) and is injected into a squid giant axon. The volume injected is of the order of 1 mm³. In the case of aminonaphthalene sulfonate derivatives, the probe concentration in the injection fluid is between 0.2 and 1 mg/ml. In some cases, a fine suspension of particles of poorly water-soluble probe molecules is introduced into an axon after removal of the major portion of the axoplasm.

Extracellular application of fluorescent probes is carried out by immersing crab nerves or squid giant axons in artificial seawater containing probe molecules. After immersion for a period of $\frac{1}{4}$–3 h, the nerve preparations are often rinsed with fresh saline solution and used for subsequent optical measurements. In other cases, optical measurements are carried out with the nerve preparations still immersed in artificial seawater containing probes. Many derivatives of aminonaphthalene sulfonate are known to be practically nonfluorescent in water and strongly fluorescent when they are bound to biological macromolecules (see Weber, 1972; Brand *et al.*, 1971; McClure and Edelman, 1966). In such cases, optical measurements are not affected by the presence of probe molecules in the surrounding medium. However, when the concentration of the probe in the medium is very high, there is absorption of the incident (exciting) light by the probes. It is therefore very important to know both the absorption and emission spectra of the probe to be tested.

Many of the probes which give rise to fluorescence signals are sensitive

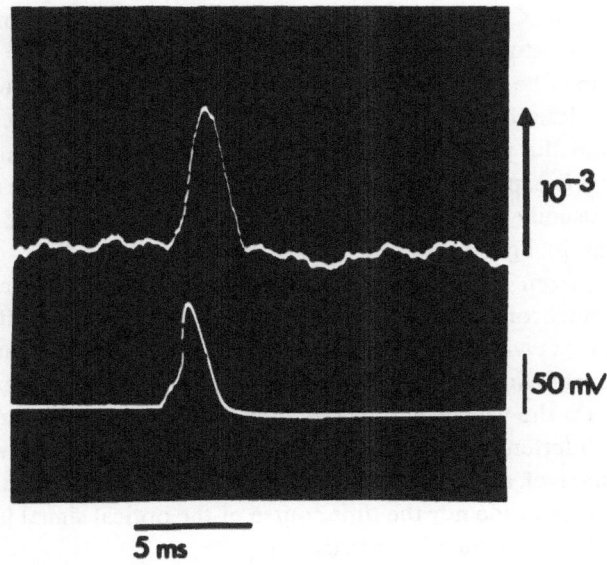

Fig. 8. Simultaneous measurement of fluorescence intensity (upper trace) and membrane potential (lower trace). A squid giant axon stained externally with merocyanine 540 (Eastman Kodak Co.) was used. From Davila *et al.* (1973).

to the solvent polarity (Tasaki *et al.*, 1973*b*). Hallett *et al.* (1972) used Ca^{2+}-sensitive probes. 4-Methylumbelliferone and other pH-sensitive probes also produce fluorescence signals. There seems no common physical basis for production of signals with different probes. Furthermore, the probe binding sites are in general very different for different probes. Probe molecules are expected to "report" to us the physicochemical properties of specific binding sites. However, since more than one type of binding site exists for each probe, physicochemical analyses of fluorescence signals are quite complicated (Tasaki *et al.*, 1973*a*, 1974).

A different type of application of the fluorescent technique has yielded some interesting results. Davila *et al.* (1973) found that some merocyanine dyes produce very large fluorescence signals. Figure 8 shows an example of the records obtained. The objective of these studies is to trace the pathway of nerve impulses by observing the fluorescent light (Salzberg *et al.*, 1973). A list of the dyes examined by these investigators may be found in Cohen *et al.* (1974).

6. LIGHT-SCATTERING SIGNALS

The optical arrangement used for studying turbidity changes during nerve excitation (Cohen *et al.*, 1968; Tasaki *et al.*, 1968) is similar to that for fluorescence studies. An interference filter (Bausch & Lomb or Infrared Industries) may be used at position F_1 in Fig. 7 to obtain monochromatic light of the desired wavelength. No optical filter (F_2) is placed between the nerve and the photomultiplier tube. A 3- to 4-mm portion of the unstained nerve is exposed to the monochromatic light and the scattered light is detected usually at 90°.

Changes in light scattering during excitation have been observed in nerve trunks from lobster and crab, as well as in squid giant axons. With 550-nm monochromatic light, the change in the light intensity scattered by crab nerves is within a range of $0.8–5.7 \times 10^{-5}$ times the intensity at rest.

The dependence of the change in light scattering during excitation of crab nerves on the wavelength of the light employed can be examined with a series of interference filters. Distinct signals are observed with monochromatic light of all the wavelengths examined between 365 and 650 nm. Neither the magnitude nor the time course of the optical signal is drastically affected by the difference in wavelength.

Applying the voltage-clamp technique to squid giant axons, Cohen *et al.* (1972*a,b*) divided a light-scattering signal into a voltage-dependent component and a current-dependent component. At present, the molecular bases of the processes contributing to these components are not well understood.

7. BIREFRINGENCE SIGNALS

In 1968, through the work of Cohen, Keynes, and Hille, it became possible to analyze conformational states of the membrane macromolecules by measuring the birefringence of the nerve (see also Tasaki *et al.*, 1968; Berestovsky *et al.*, 1969). The principle of the method of detecting birefringence signals is shown in Fig. 9 (top). When a beam of quasi-monochromatic light passes through a polarizer, a linearly polarized light beam is obtained. In the absence of a nerve in the light pathway, propagation of this polarized light is completely blocked by an analyzer placed in the crossed position. Insertion of a birefringent material such as a nerve between these two Polaroid sheets makes the light reach the detector again. The intensity of this light is maximal when the long axis of the nerve is at approximately 45° to both of the polarizing axes of the sheets. This maximum intensity, *I*, is related to the refractive indices of the nerve in the longitudinal and transverse directions, *n* and *n'*, respectively:

$$I = I_0 \sin^2 \frac{\pi d(n - n')}{\lambda}$$

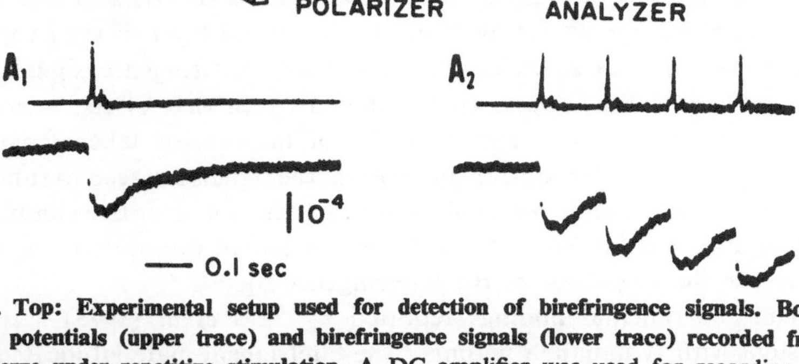

Fig. 9. Top: Experimental setup used for detection of birefringence signals. Bottom: Action potentials (upper trace) and birefringence signals (lower trace) recorded from a crab nerve under repetitive stimulation. A DC amplifier was used for recording these optical signals. Note that birefringence signals summate. From Watanabe *et al.* (1973).

where I_0 represents the intensity of the incident light, λ the wavelength, and d the thickness of the nervous tissue (e.g., see p. 247 in Born, 1933). A birefringence signal represents a transient decrease in I during nerve excitation associated with a reduction in the difference between the two refractive indices.

It is important to note that the effect of light scattering and absorption is completely ignored in the mathematical formula quoted above. In crab nerves, the loss of light due to scattering is quite significant (Watanabe and Terayama, unpublished). However, it has been shown that a transient decrease in the observed light intensity represents a true birefringence signal (Cohen et al., 1968; Tasaki et al., 1968).

The polarizers and analyzers used in these experiments are inexpensive Polaroid sheets (HN38 or KN36) which yield quite satisfactory results. A detector of the semiconductor type (e.g., Photodiode, United Detector Technology, Inc., Santa Monica, Calif.) can be used to observe small transient changes in light intensity. In crab nerves, birefringence signals are known to be large enough to be seen directly on an oscillograph screen without using a signal averager (Tasaki et al., 1968). Von Muralt (1971) found that olfactory nerves of garfish give rise to even larger birefringence signals.

There are now several different interpretations of the data obtained. Berestovsky et al. (1970) asserted that the birefringence signals are due to the molecular mechanism controlling the electrical conductivity of the membrane. Cohen et al. (1971) maintained that the signals are due to a Kerr effect produced by the potential variation associated with an action potential and are not related to the membrane conductivity. The interpretation proposed in this laboratory is that these signals arise from a transient decrease in birefringence originating from the longitudinally oriented fibrous material found near the axonal membrane (Sato et al., 1973).

Quite recently, Watanabe et al. (1973) have shown that the major portion of these signals derives from the superficial layer of the axoplasm. When recorded with a DC-coupled amplifier, birefringence signals were found to have a time course quite different from that of the membrane potential. Figure 9 shows two examples of the records taken from crab nerves. In response to repetitive stimulation, the signals are seen to summate. As is well known, action potentials never summate on repetitive stimulation. Furthermore, a reduction in the external Ca ion concentration was found to decrease the amplitude of the birefringence signals.

The experimental findings reported by Sato et al. (1973) are quite consistent with Watanabe's findings. The birefringent material located near the axonal membrane was shown to be digestible with pronase.

8. CONCLUSIONS

In this chapter, we have described the technique of preparing several kinds of excitable tissue and nerve fibers commonly used for studying electrophysiological properties of the plasma membrane, the method of intracellular perfusion, and various optical methods for detecting signs of conformational changes in the membrane macromolecules. Descriptions of the method of microelectrode recording and of the voltage-clamp technique were omitted in this chapter because a large number of articles and reviews can be found on these methods.

As in other branches of science, discovery of a new method opens up a new field of investigation which leads us to modify or abandon old concepts. Or, rather, significant progress in our understanding of the process of nerve excitation is brought about only by discovery of new methods. The method of intracellular perfusion and the optical methods are the most recent additions to the arsenal of investigators in this field.

At present, many attempts are being made to discover new non-electrical means of studying the process of excitation in the nerve membrane. The major difficulties for some investigators seem to derive from their unfamiliarity (1) with nerve preparations commonly used in electrophysiological studies and (2) with the rapidity and smallness of the desired signals. It is hoped that this chapter will be of some help in developing new instruments which will enable us to follow rapid physicochemical processes occurring in the nerve membrane during excitation.

ACKNOWLEDGMENT

The authors are deeply indebted to Mrs. Yuri Inoue, who prepared the illustrations for this chapter.

9. REFERENCES

Adrian, E. D., and Bronk, D. W., 1928, The discharge of impulses in motor nerve fibres. I. Impulses in single fibres of the phrenic nerve, *J. Physiol. (London)* **66**:81.

Arnold, J. M., Gilbert, D. L., Daw, N. W., Summers, W. C., Manalis, R. S., and Lasek, R. J., 1974, *A Guide to Laboratory Use of the Squid Loligo pealei*, Marine Biological Laboratory, Woods Hole, Mass.

Baker, P. F., Hodgkin, A. L., and Shaw, T. I., 1961, Replacement of protoplasm of a giant nerve fibre with artificial solutions, *Nature* **190**:885.

Baker, P. F., Hodgkin, A. L., and Shaw, T. I., 1962, The effects of changes in internal ionic concentrations on the electrical properties of perfused giant axons, *J. Physiol.* (*London*) **164**:355.

Berestovsky, G. N., Lunevsky, V. Z., Musienko, V. S., and Razhin, V. D., 1969, Rapid changes in birefringence of the nerve fiber membrane, *Dokl. Acad. Nauk SSSR* **189**:203.

Berestovsky, G. N., Frank, G. M., Liberman, E. A., Lunevsky, V. Z., and Razhin, V. D., 1970, Electrooptical phenomena in bimolecular phospholipid membranes, *Biochim. Biophys. Acta* **219**:263.

Bernstein, J., 1902, Untersuchungen zur Thermodynamik der bioelektrischen Ströme, *Pfluegers Arch. Gesamte Physiol.* **92**:521.

Binstock, L., and Goldman, L., 1969, Current- and voltage-clamped studies on *Myxicola* giant axons, *J. Gen. Physiol.* **54**:730.

Born, M., 1933, *Optik*, Springer, Berlin.

Brand, L., and Gohlke, J. R., 1972, Fluorescence probes for structure, *Ann. Rev. Biochem.* **41**:843.

Brand, L., Seliskar, C. J., and Turner, D. C., 1971, The effects of chemical environment on fluorescent probes, in: *Probes of Structure and Function of Macromolecules and Membranes*, Vol. 1 (B. Chance, C.-P. Lee, and J. K. Blasie, eds.), p. 17, Academic Press, New York.

Cavanaugh, G. M. (ed.), 1956, *Marine Biological Laboratory: Formulae and Methods*, Woods Hole, Mass.

Cohen, L., Keynes, R. D., and Hille, E., 1968, Light scattering and birefringence during nerve activity, *Nature* **218**:433.

Cohen, L. B., Hille, B., Keynes, R. D., Landowne, D., and Rojas, E., 1971, Analysis of the potential-dependent changes in optical retardation in the squid giant axon, *J. Physiol.* (*London*) **218**:205.

Cohen, L. B., Keynes, R. D., and Landowne, D., 1972a, Changes in light scattering that accompany the action potential in squid giant axons: Potential-dependent components, *J. Physiol.* (*London*) **224**:701.

Cohen, L. B., Keynes, R. D., and Landowne, D., 1972b, Changes in axon light scattering that accompany the action potential: Current-dependent components, *J. Physiol.* (*London*) **224**:727.

Cohen, L. B., Salzberg, B. M., Davila, H. V., Ross, W. N., Landowne, D., Waggoner, A. S., and Wang, C.-H., 1974, Changes in axon fluorescence during activity: A search for useful probes, *J. Membr. Biol.* **19**:1.

Davies, P. W., 1961, A method for measuring membrane potential of intracellularly perfused single skeletal muscle fibers, *Fed. Proc.* **20**:142.

Davila, H. V., Salzberg, B. M., Cohen, L. B., and Waggoner, A. S., 1973, A large change in axon fluorescence that provides a promising method of measuring membrane potential, *Nature New Biol.* **241**:159.

Fishman, H. M., 1970, Direct and rapid description of the individual ionic currents of squid axon membrane by ramp potential control, *Biophys. J.* **10**:799.

Frankenhaeuser, B., 1957, A method for recording resting and action potentials in the isolated myelinated nerve fibre of the frog, *J. Physiol.* (*London*) **135**:550.

Furusawa, K., 1929, The depolarization of crustacean nerve by stimulation or oxygen want, *J. Physiol.* (*London*) **67**:325.

Hallett, M., Schneider, A. S., and Carbone, E., 1972, Tetracycline fluorescence as calcium-probe for nerve membrane with some model studies using erythrocyte ghosts, *J. Membr. Biol.* **10**:31.

Hartline, H., 1938, The response of single optic nerve fibers of the vertebrate eye to illumination of the retina, *Am. J. Physiol.* **121**:400.

Inoue, I., Ishima, Y., Horie, H., and Takenaka, T., 1971, Properties of excitable membrane produced on the surface of protoplasmic drop in *Nitella*, *Proc. Jp. Acad.* **47**:549.

Inoue, I., Ueda, T., and Kobatake, Y., 1973, Structure of excitable membranes formed on the surface of protoplasmic drops isolated from *Nitella*. I. Conformation of surface membrane determined from the refractive index and from enzyme actions, *Biochim. Biophys. Acta* **298**:653.

Inoue, I., Tasaki, I., and Kobatake, Y., 1974, A study of the effects of externally applied sodium-ions and detection of spatial non-uniformity of the squid axon membrane under internal perfusion, *Biophys. Chem.* **2**:116.

Johnson, F. H., and Haneda, Y., 1966, *Bioluminescence in Progress*, pp. 35–52, Princeton University Press, Princeton, N.J.

Kato, G., 1934, *The Microphysiology of Nerve*, Maruzen, Tokyo.

Ling, G. N., 1960, The interpretation of selective ionic permeability and cellular potentials in terms of the fixed charge-induction hypothesis, *J. Gen. Physiol. Suppl.* **43**:149.

McClure, W. O., and Edelman, G. M., 1966, Fluorescent probes for conformational states of proteins. I. Mechanism of fluorescence of 2-*p*-toluidinylnaphthalene-6-sulfonate, a hydrophobic probe, *Biochemistry* **5**:1908.

Oikawa, T., Spyropoulos, C. S., Tasaki, I., and Teorell, T., 1961, Methods for perfusing the giant axon of *Loligo pealei*, *Acta Physiol. Scand.* **52**:195.

Osterhout, W., and Hills, S., 1938, Calculations of bioelectric potentials, *J. Gen. Physiol.* **22**:541.

Passo, H., and Stämpfli, R. (ed.), 1969, *Laboratory Techniques in Membrane Biophysics*, Springer, Berlin.

Radda, G. K., and Vanderkooi, J., 1972, Can fluorescent probes tell us anything about membranes? *Biochim. Biophys. Acta* **265**:509.

Robinson, R. A., and Stokes, R. H., 1959, *Electrolyte Solutions*, 2nd ed., Butterworth, London.

Salzberg, B. M., Davila, H. V., and Cohen, L. B., 1973, Optical recording of impulses in individual neurones of an invertebrate central nervous system, *Nature* **246**:508.

Sato, H., Tasaki, I., Carbone, E., and Hallett, M., 1973, Changes in axon birefringence associated with excitation: Implications for the structure of the axon membrane, *J. Mechanochem. Cell Motil.* **2**:209.

Shrager, P., Macey, R., and Strickholm, A., 1969, Internal perfusion of crayfish giant axons: Action of tannic acid, DDT, and TEA, *J. Cell. Physiol.* **74(1)**:77.

Stämpfli, R., 1959, Is the resting potential of Ranvier nodes a potassium potential? *Ann. N.Y. Acad. Sci.* **81**:265.

Stryer, L., 1968, Fluorescence spectroscopy of proteins, *Science* **162**:526.

Tasaki, I., 1939, The strength–duration relation of the normal polarized and narcotized nerve fiber, *Am. J. Physiol.* **125**:367.

Tasaki, I., 1953, *Nervous Transmission*, Charles C. Thomas, Springfield, Ill.

Tasaki, I., 1968, *Nerve Excitation: A Macromolecular Approach*, Charles C. Thomas, Springfield, Ill.

Tasaki, I., and Frank, K., 1955, A measurement of the action potential of myelinated nerve fiber, *Am. J. Physiol.* **182**:572.

Tasaki, I., and Singer, I., 1966, Membrane macromolecules and nerve excitability: A physico-chemical interpretation of excitation in squid giant axons, *Ann. N.Y. Acad. Sci.* **137**:792.

Tasaki, I., Watanabe, A., and Takenaka, T., 1962, Resting and action potential of intracellularly perfused squid giant axons, *Proc. Natl. Acad. Sci. (USA)* **48**:1177.

Tasaki, I., Singer, I., and Takenaka, T., 1965, Effects of internal and external ionic environment on excitability of squid giant axon, *J. Gen. Physiol.* **48**:1095.

Tasaki, I., Watanabe, A., Sandlin, R., and Carnay, L., 1968, Changes in fluorescence turbidity and birefringence associated with nerve excitation, *Proc. Natl. Acad. Sci. (USA)* **61**:883.

Tasaki, I., Carnay, L., Sandlin, R., and Watanabe, A., 1969a, Fluorescence changes during conduction in nerves stained with acridine orange, *Science* **163**:683.

Tasaki, I., Carnay, L., and Watanabe, A., 1969b, Transient changes in extrinsic fluorescence of nerve produced by electric stimulation, *Proc. Natl. Acad. Sci. (USA)* **64**:1362.

Tasaki, I., Watanabe, A., and Hallett, M., 1972, Fluorescence of squid axon membrane labelled with hydrophobic probes, *J. Membr. Biol.* **8**:109.

Tasaki, I., Carbone, E., Sisco, K., and Singer, I., 1973a, Spectral analyses of extrinsic fluorescence of the nerve membrane labeled with aminonaphthalene derivatives, *Biochim. Biophys. Acta* **323**:220.

Tasaki, I., Hallett, M., and Carbone, E., 1973b, Further studies of nerve membranes labeled with fluorescent probes, *J. Membr. Biol.* **11**:353.

Tasaki, I., Sisco, K., and Warashina, A., 1974, Alignment of anilinonaphthalenesulfonate and related fluorescent probe molecules in squid axon membrane and in synthetic polymers, *Biophys. Chem.*, **2**:316.

Teorell, T., 1953, Transport processes and electrical phenomena in ionic membranes, *Progr. Biophys.* **3**:305.

von Muralt, A., 1971, "Optical spike" during excitation in peripheral nerve, p. 638, Abst. 25th Int. Physiol. Congr., Munich.

Watanabe, A., Terayama, S., and Nagano, M., 1973, Axoplasmic origin of the birefringence change associated with excitation of a crab nerve, *Proc. Jpn. Acad.* **49**:470.

Weber, G., 1972, Uses of fluorescence in biophysics: Some recent developments, *Ann. Rev. Biophys. Bioeng.* **1**:553.

Index